STUDENT UNIT GUIDE

NEW EDITION

Edexcel AS Geography Unit 2
Geographical Investigations

David Holmes and Bob Hordern

PHILIP ALLAN

Philip Allan Updates, an imprint of Hodder Education, an Hachette UK company, Market Place, Deddington, Oxfordshire OX15 0SE

Orders
Bookpoint Ltd, 130 Milton Park, Abingdon, Oxfordshire, OX14 4SB
tel: 01235 827827
fax: 01235 400401
e-mail: education@bookpoint.co.uk
Lines are open 9.00 a.m.–5.00 p.m., Monday to Saturday, with a 24-hour message answering service.
You can also order through the Philip Allan Updates website: www.philipallan.co.uk

ISBN 978-1-4441-4764-3

Impression number 5 4 3 2 1
Year 2015 2014 2013 2012 2011

Printed in Dubai

Hachette UK's policy is to use papers that are natural, renewable and recyclable products and made from wood grown in sustainable forests. The logging and manufacturing processes are expected to conform to the environmental regulations of the country of origin.

P01910

Contents

Getting the most from this book

Examiner tips

Advice from the examiner on key points in the text to help you learn and recall unit content, avoid pitfalls, and polish your exam technique in order to boost your grade.

Knowledge check

Rapid-fire questions throughout the Content Guidance section to check your understanding.

Knowledge check answers

1 Turn to the back of the book for the Knowledge check answers.

Summary

Summaries

- Each core topic is rounded off by a bullet-list summary for quick-check reference of what you need to know.

Questions & Answers

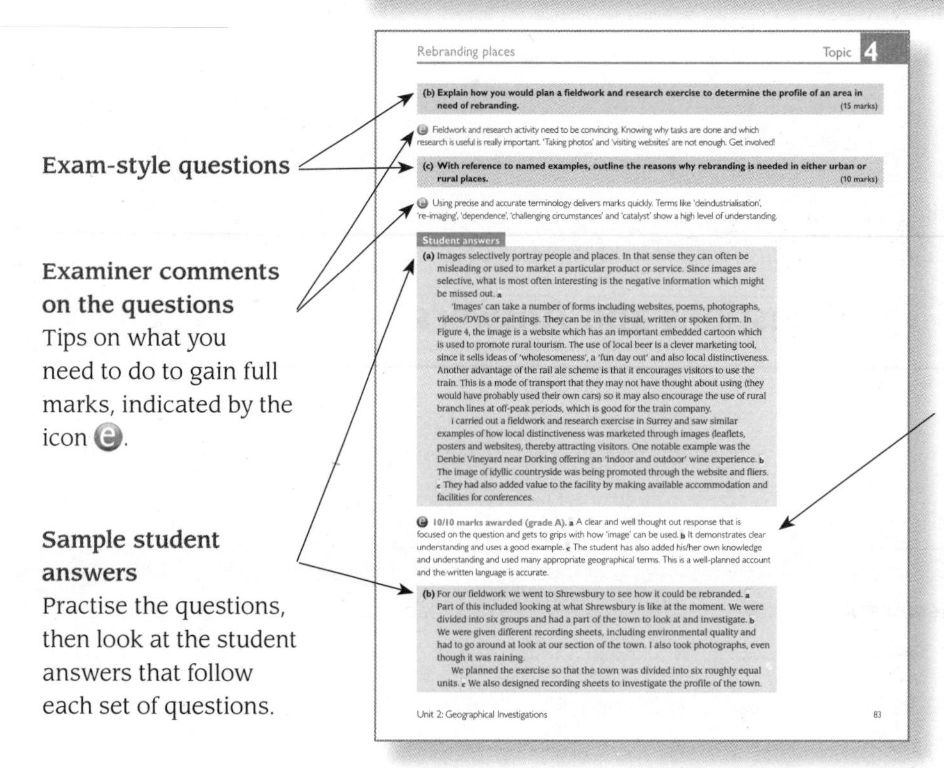

Rebranding places — Topic 4

(b) Explain how you would plan a fieldwork and research exercise to determine the profile of an area in need of rebranding. (15 marks)

(e) Fieldwork and research activity need to be convincing. Knowing why tasks are done and which research is useful is really important. 'Taking photos' and 'visiting websites' are not enough. Get involved!

(c) With reference to named examples, outline the reasons why rebranding is needed in either urban or rural places. (10 marks)

(e) Using precise and accurate terminology delivers marks quickly. Terms like 'deindustrialisation', 're-imaging', 'dependence', 'challenging circumstances' and 'catalyst' show a high level of understanding.

Student answers

(a) Images selectively portray people and places. In that sense they can often be misleading or used to market a particular product or service. Since images are selective, what is most often interesting is the negative information which might be missed out. **a**

'Images' can take a number of forms including websites, poems, photographs, videos/DVDs or paintings. They can be in the visual, written or spoken form. In Figure 4, the image is a website which has an important embedded cartoon which is used to promote rural tourism. The use of local beer is a clever marketing tool, since it sells ideas of 'wholesomeness', a 'fun day out' and also local distinctiveness. Another advantage of the rail ale scheme is that it encourages visitors to use the train. This is a mode of transport that they may not have thought about using (they would have probably used their own cars) so it may also encourage the use of rural branch lines at off-peak periods, which is good for the train company.

I carried out a fieldwork and research exercise in Surrey and saw similar examples of how local distinctiveness was marketed through images (leaflets, posters and websites), thereby attracting visitors. One notable example was the Denbie Vineyard near Dorking offering an 'indoor and outdoor' wine experience. **b** The image of idyllic countryside was being promoted through the website and fliers. **c** They had also added value to the facility by making available accommodation and facilities for conferences.

(e) **10/10 marks awarded (grade A).** **a** A clear and well thought out response that is focused on the question and gets to grips with how 'image' can be used. **b** It demonstrates clear understanding and uses a good example. **c** The student has also added his/her own knowledge and understanding and used many appropriate geographical terms. This is a well-planned account and the written language is accurate.

(b) For our fieldwork we went to Shrewsbury to see how it could be rebranded. **a** Part of this included looking at what Shrewsbury is like at the moment. We were divided into six groups and had a part of the town to look at and investigate. **b** We were given different recording sheets, including environmental quality and had to go around at look at our section of the town. I also took photographs, even though it was raining.

We planned the exercise so that the town was divided into six roughly equal units. **c** We also designed recording sheets to investigate the profile of the town.

Unit 2: Geographical Investigations 83

Exam-style questions

Examiner comments on the questions
Tips on what you need to do to gain full marks, indicated by the icon (e).

Sample student answers
Practise the questions, then look at the student answers that follow each set of questions.

Examiner commentary on sample student answers
Find out how many marks each answer would be awarded in the exam and then read the examiner comments (preceded by the icon (e)) following each student answer. Annotations that link back to points made in the student answers show exactly how and where marks are gained or lost.

About this book

The **Content Guidance** section of this guide identifies the bare essentials of Unit 2 and is split into its four topics — Extreme weather, Crowded coasts, Unequal spaces, and Rebranding places. Each main topic is divided into four sub-sections. Exam guidance in the form of flow diagrams is included after each sub-section. This links the content of the specification to fieldwork and research, including planning, methodology, presentation and analysis.

The **Question & Answer** section includes sample questions and answers on each of the four topics plus comments on how questions should have been tackled, and where marks could have been gained or lost.

Fieldwork and research

Although some of the questions that make up Unit 2 test the knowledge and understanding gained by following the course, fieldwork and research form important components of the Edexcel A-level specification and encompass a number of important ideas. New technologies, including GIS and electronic maps, are particularly important, as is the use of the internet to access additional research information.

The Edexcel specification offers exciting and original opportunities in terms of fieldwork and research, as there is no requirement to complete a formal write-up. Innovative ways of following up the investigative process can be used, and statistics can sometimes be appropriate when dealing with quantitative data. Statistics, though, should not dominate the fieldwork approach or influence decisions on how data are collected.

Many qualitative approaches lend themselves to data analysis methods that could involve the use of annotated photographs, documents, personalised maps, blogs, movies/DVDs, geographical records etc.

Being a hunter-gatherer

A good hunter-gatherer squirrels away resources for use at a later date. It is never too early to collect information that may be relevant to fieldwork or research. An exam could ask you to discuss how and why you collected information, or to comment on resources provided. Information can come from specialist books and magazines and from newspapers or journals (paper-based or online versions).

Effective searching on the internet

A wide range of material can be gathered from the internet. A number of reliable search engines can be used. They all provide slightly different information, and you are advised to check more than one.

Students sometimes fail to use the internet effectively because of:

- a lack of search strategy and too much trial and error
- a tendency to stick with what is known rather than what is appropriate
- little understanding of advanced functions of different search engines
- an expectation that searches will answer questions fully
- poor judgement of what is relevant, factually accurate, or up to date
- a lack of critical and evaluative skills
- snap decisions on usefulness

The quality of internet information can vary greatly and pages and whole sites can move or disappear without warning. Skill is needed to retrieve what is really required.

Internet researchers should consider:

- who is responsible for the document?
- are they credible authors?
- can sources be checked?
- is there editorial input indicating quality control?
- are any biases and affiliations clearly stated?
- are authors' opinions clearly identified as such?
- does the document have a date, and how up to date is it?

Content guidance

Unit 2: Geographical Investigations comprises four topics:

1 Extreme weather
2 Crowded coasts
3 Unequal spaces
4 Rebranding places

You are required to study two topics — 1 or 2 and 3 or 4. Fieldwork and research are integral to the investigation process.

In this **Content Guidance** section each topic follows the sequence identified in the specification — a description of the issue/problem and its context, impacts and risks, and how problems can be managed. Urban or rural environments could be specified.

Fieldwork and research opportunities are identified in each section, along with key terms in **bold**. You should use these terms to create an ongoing glossary as you work through the book. Some case studies are included; many of these are UK-based, although examples from other parts of the world can be equally relevant.

Extreme weather

Extreme weather watch

What do we mean by 'extreme weather'?

Extreme weather can be described as severe, unexpected or at record levels. The severity scale ranges from major hurricanes to localised flash floods. Extreme temperatures can mean cold snaps or heatwaves, while sudden downpours and drought illustrate the consequences of extreme rainfall events.

Knowledge check 1

Define the term 'extreme weather'.

Conditions can be classified as **immediate** (a tornado), **consequential** (floods) or **long-term** (drought). Some examples of extreme weather are given below:

- The 2005 hurricane season in the USA was extreme, with 249 storms and 13 hurricanes — including three of the worst on record.
- The La Niña event of 2010/11 brought cyclones (e.g. Tasha) and record rainfall to Queensland, Australia, flooding almost half the state.
- The driest place on earth is Quillagua, Chile, with only 0.05 mm of rain per year. Vostok, in Antarctica, is the coldest (–89°C) and windiest location.

Examiner tip

It is important to spread your work across a number of different types of extreme weather.

Investigating extreme weather

This is a core fieldwork and research opportunity, in which you can observe and monitor changes in weather conditions and carry out research into UK weather systems and their underlying meteorology.

Fieldwork can involve weather recording or keeping a diary of weather conditions and related events. This could be linked to the effects of climate change, variations in UK air masses, and the arrival of weather systems such as depressions and anticyclones. Primary data could be recorded using a traditional weather station, making daily observations, or using a continuous electronic data logger. Keeping a weather diary over even a short period can help you understand how the weather changes and why events happen. Take a look at 'Practical geography: local weather studies' in *Geography Review*, Vol. 20, No. 1.

Weather research should involve looking at contrasting weather systems, as these can, potentially, lead to extreme weather conditions. Comparing phases of UK weather can be rewarding, and monitoring how weather affects local transport, watcr supplies and sports events is useful for the next section on weather impacts. The Meteorological Office website (**www.metoffice.gov.uk/weather/uk**) is a good place to start researching weather. Alternatively, visit **www.bbc.co.uk/weather**.

Case Study

UK depressions and severe weather

Depressions are low-pressure systems that form when a strong jet stream in the troposphere brings tropical maritime air into contact with a polar air mass. This forms a vortex of unstable rising air along the polar front, triggering rapid temperature changes, strong winds and pulses of rain. The characteristic sequence involved is shown in the cross-section (Figure 1). The sequence should be read from right to left. The same features can be seen on synoptic maps, which use tightly packed isobars and various symbols to show the severity of the storm involved.

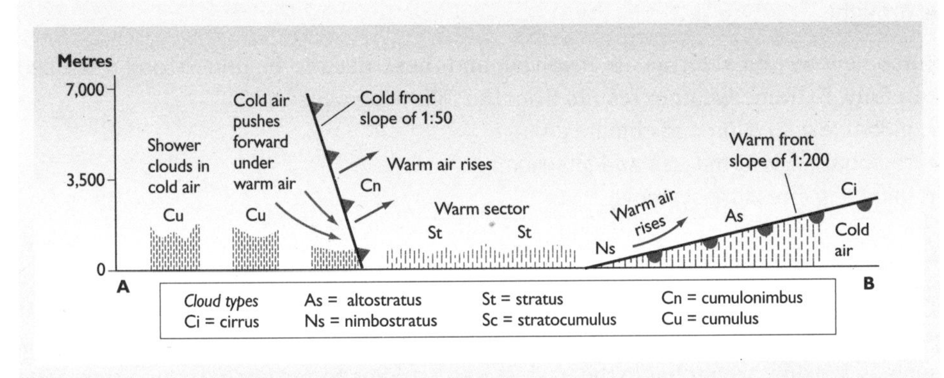

Figure 1 How weather changes as a depression passes

Cyclonic weather prevails for 20% of the year over the UK and can occasionally lead to severe weather conditions — as in the summer floods of 2007. Some examples of severe weather conditions resulting from mid-latitude depressions are shown in Table 1.

Table 1 Depressions and severe weather events in the UK

Event, date and situation	Features	Impacts
Great Storm, October 1987 Air from Hurricane Floyd over the Bay of Biscay meets cold front	Very low air pressure, with winds of hurricane force	12 deaths; 14 million trees lost; ferries ran aground; £2 billion insurance bill
Towyn surge, February 1990 Deep depression, gale-force winds and high spring tide	Strong on-shore winds, and massive waves	Towyn sea wall and railway embankment breached; extensive flooding; 2,800 properties damaged
Boscastle floods, August 2004 Heavy rainfall over already saturated steep-sided valleys	Torrential rain (88 mm in 1 hour) caused flash floods	Property damage to most premises in Boscastle and emergency evacuation of 120 people required; no deaths
Severn floods, July 2007 Deflected (southerly) jetstream led to a continuing series of depressions	June rainfall (134 mm) was twice average; July rain (234 mm) was four times average	Significant property damage; 250,000 people without clean water; insurance bill over £3.25 billion

Knowledge check 2

Identify those features of a depression (Table 1) that can cause hazardous conditions.

How does extreme weather develop?

The final item in this section is about causes and events. You should have some knowledge of how the three types of extreme weather listed below can develop, though for revision purposes it may be useful to focus on one:

- hurricanes
- winter conditions
- drought

Examiner tip

Be an expert in at least two of these types of extreme weather: hurricane, drought, depressions or winter weather.

Some new technical terms are involved, and these need to be understood and used carefully. Extreme weather results from the following causes:

- global (e.g. El Niño and climate change)
- regional (e.g. air masses and jetstreams)
- local (e.g. pressure systems)

Hurricane development

The way in which hurricanes develop and migrate shows how meteorological processes and weather systems are inter-related. Atlantic or Caribbean hurricanes such as Katrina are probably the easiest case studies to research, though typhoons and cyclones do make useful comparisons.

Knowledge check 3

Focusing on hurricanes, very briefly explain:

- their origins (location and causes)
- their characteristics (features and development)

Hurricanes are storms that originate in the tropics and reach sustained wind speeds of over 120 km per hour. Most form in the low-pressure area north of the equator between July and October, where the trade winds meet over the warm ocean. In this inter-tropical convergence zone (ITCZ) sea temperatures are above 26°C and air humidity is over 75%. As unstable, moist air rises it triggers thunderstorms which group together to form a high-energy hurricane (see Figure 2). During the Atlantic hurricane season, from 15 May to 30 November, hurricanes move northwestwards, gathering strength and speed.

- Falling air pressure creates an upward spiral of increasing wind speeds.
- Water vapour rises, cools rapidly and creates a wall of cloud around the 'eye'.
- Condensation increases wind speeds, pushing rising air up to 10 km in altitude.
- At the top, cooling air spreads a thick canopy of cirrus clouds up to 1,500 km across.
- Strong winds, torrential rain and massive storm waves are produced.

Examiner tip
Practise drawing a sketch of Figure 2 and getting the labels correct.

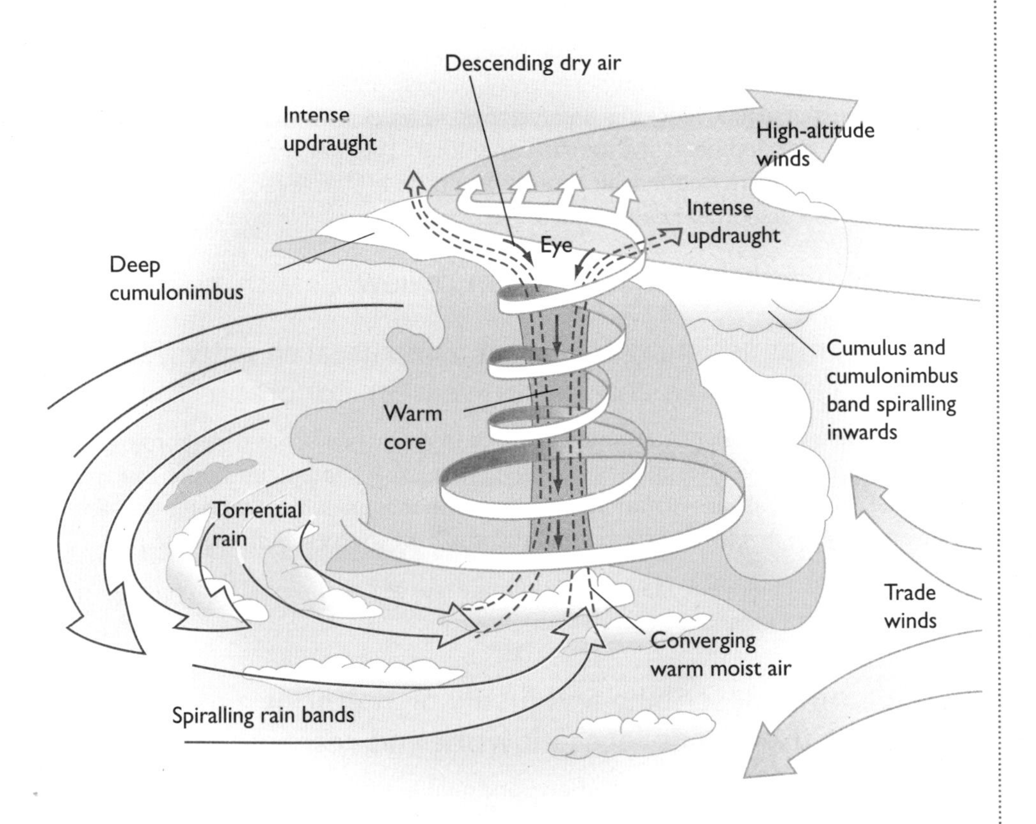

Figure 2 Anatomy of a hurricane

Winter conditions

These are essentially about snow and ice. The two examples below are both well documented.

- A major storm in March 1993 brought hurricane-force winds and blizzards to much of the east coast of the USA and temperatures dropped to –10°C. This low pressure (960 mb) was caused when the polar jetstream mixed moist tropical air from the Gulf with cold northeasterly air. Road, rail and air travel was halted, 130 million people in 26 states were affected and 300 people died.
- The winter of 1947 remains the UK's snowiest on record, with prolonged disruption including travel chaos, food shortages and hypothermia. A large anticyclone of polar continental air from Scandinavia brought heavy snowfall and freezing temperatures. Then, in March 1948, the arrival of Atlantic depressions triggered widespread flooding as the snow melted.

Examiner tip

Extreme weather patterns vary from year to year. Look at the graphs on Figure 1 in the Q&A section (p. 72) to confirm this.

Knowledge check 4

Explain why most types of extreme weather seem to occur seasonally.

Drought

The concept of drought needs to be approached carefully. The tropical version seen in parts of Africa is considerably more extreme than its counterpart in southeast England.

- In the Sahel belt of Africa, rainfall is low and unpredictable, making it the most drought-prone area of the world. The repeated failure of the rains at the end of the dry season brings extreme conditions to countries like Ethiopia, Sudan and Chad. Human actions such as deforestation, overgrazing and collecting firewood could be contributory factors, leading to poverty, disease and conflict.
- In the UK, persistent blocking anticyclones — as seen in summer 2003 — can lead to the development of drought conditions. This pattern of weather, which could become more frequent in future, results in a high level of water demand, hyperthermia and other health issues.

Extreme weather watch — exam review

Key outcome: designing fieldwork and research activities to monitor and observe changes in weather conditions

It is important in weather-related studies that changes in atmospheric meteorological conditions are observed for at least 1 week. This work should contribute towards a weather diary. Discuss locations of observations, including advantages and disadvantages of sites (considering aspects, proximity to trees, buildings etc.).

Other fieldwork could include a microclimate survey to investigate changes in weather in a small area.

Key outcome: describing and justifying methods and techniques used to collect fieldwork and research data

Fieldwork could involve a mix of qualitative and quantitative approaches: **Qualitative** methods include field notes, field sketches and photographs (of sky, site, equipment, cloud cover etc.).

Quantitative methods could include weather measurements (relative humidity, wind speed and direction, rainfall, maximum and minimum temperatures, visibility).

Use weather-specific websites to provide satellite imagery (rainfall maps, forecasts and site-specific weather data) or use observations from the Met Office. Weather reports could also be derived from television, newspapers and radio. Compare the reliability of each.

Key outcome: describing and justifying techniques chosen to present and analyse findings

Choice and range of presentation techniques will be influenced by data type, such as a time-lapse DVD of sky changes captured using a webcam. An accompanying transcript or narrative could form part of the analysis, and clouds could be recorded as simple sketches or pie charts. Other data could be processed, tabulated and presented using bar charts of rainfall, temperature, pressure and humidity, plus rose diagrams of wind speed and direction.

Analysis using simple statistics could also be appropriate (mode, mean and median of observed weather recordings, providing there are sufficient data to make analysis meaningful). Students could also use annotated satellite imagery, including a discussion of underlying meteorological conditions, flow diagrams and mind-maps of observed changes in atmosphere over a period of time.

Key outcome: commenting on the data and conclusions; evaluating limitations

Summarise data patterns, trends and anomalies as revealed by the analysis of data types, both observed and obtained from other sources. Provide a summary of the fieldwork process (how does weather vary from time to time and under different background synoptic conditions?). How does weather vary within a small area?

How reliable are the results and what factors could have influenced reliability of weather data (poor site, unreliable equipment etc.)? Compare directly observed results with those published from other sources.

If the design process was to be modified, how could it be changed in terms of locations, range of observation techniques, methods of presentation and analysis?

Examiner tip

Be prepared to write about your:

- fieldwork activities – what you did and the results this gave you
- research – what sources you used and what they showed

Knowledge check 5

Describe what each of these could tell you:

- a weather diary or station records
- a satellite image
- a television weather map (synoptic chart)

Extreme impacts

How and why do extreme weather impacts vary?

This section is about the impacts of extreme weather, beginning with a look at how impacts can differ. Impacts are different because some types of extreme weather are larger, more common, more damaging and more deadly than others. A hazard profile (see Figure 3) is one way to compare them.

Examiner tip

Impacts of extreme weather vary considerably in size, duration and frequency (see Figure 3).

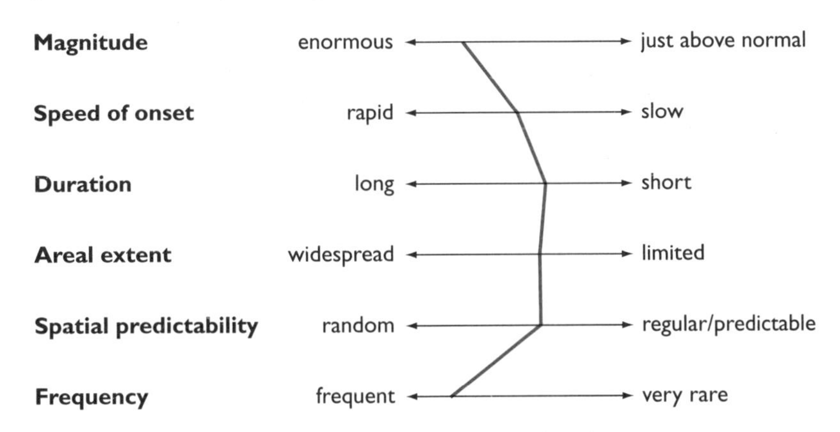

Figure 3 A hazard profile for a hurricane

Impacts relate to the severity of the event. Most hazards have a scale that calibrates magnitude and effects. Hurricanes are measured using the Saffir-Simpson scale (Table 2). Tornadoes use the Fujita scale.

Table 2 The Saffir-Simpson scale

Category	Damage	Winds (km/hour)	Storm surge (m)
1	Minimal	119–53	1.22–1.81
2	Moderate	154–77	1.82–2.73
3	Extensive	178–209	2.74–3.95
4	Extreme	210–49	3.96–5.49
5	Catastrophic	250 and over	5.5 and over

Knowledge check 6

How does the Saffir-Simpson scale work?

Investigating severe weather impacts

This is a core fieldwork and research opportunity. You should use primary and secondary sources to investigate the social and economic impacts of extreme weather (environmental effects could also be considered). This work should include three elements:

- an immediate disastrous weather event — tornado or hurricane
- a subsequent additional hazard — localised river flooding
- a longer-term trend or condition — heatwave or drought

Flood impact fieldwork can concentrate on localised flood impacts along a small stream or part of a larger catchment. This work should be planned and designed carefully. A flood impact investigation in a town could include the following:

- a land-use survey of floodplain (open space, industrial, residential etc.)
- mapping of areas vulnerable to flooding (contours and flood-defence information)
- carrying out a transect across the floodplain (plot flood recurrence levels)
- taking photographs of relevant features (bridge clearances, defences, flooding)
- using targeted questionnaires to canvas local views on risks and management
- arranging interviews with key players (e.g. local Environment Agency staff)

Examiner tip

Carry out fieldwork into the effects of flooding in a small area (riverside town).

Examiner tip

Know the sources where your research information came from.

Flood research could be used with this practical work to create a fuller understanding of flood impacts. The following sources might be explored:

- the Environment Agency website — **www.environment-agency.gov.uk**
- a flood-risk map (type the postcode into the flood section of the above site)
- river discharge data, which can often be found for a nearby gauging station
- local newspaper files of previous floods (information and photographs)

Knowledge check 7

Explain how you could use maps in fieldwork to help with a flood impact survey.

Additional research activities

Hurricanes and tornadoes illustrate the sudden impact that severe weather can have, while the effects of heatwaves and extended droughts often develop over a longer period of time. All can have significant social, economic and environmental impacts:

- Hurricanes — choices for study could be Katrina or a cyclone like Yasi.
- Tornadoes or supercells — most examples come from America's Tornado Alley, where Oklahoma is the most tornado-prone state. In the last week of April 2011, tornadoes and violent storms ripped through seven Southern states, killing at least 295 people and causing billions of dollars of damage in some of the deadliest twisters in US history. It was the worst US natural disaster since Hurricane Katrina in 2005.
- Heatwaves — recent summers have seen long periods of high pressure in parts of Europe. In August 2003, 35,000 people (mostly elderly) died in the record

temperatures. In 2007, Greece and the Balkan states faced forest fires, water supply problems, and the collapse of many business and agricultural projects.

- Droughts — these can occur almost anywhere in the world. Tropical varieties tend to be more severe and long lasting, e.g. in the Sahel. Droughts in Australia and southwest USA show how richer countries can be badly affected (see the following case study and **www.drought.gov**).

Examiner tip
The effects of drought can be direct (e.g. deaths) and indirect (e.g. food prices)

Table 3 The impacts of Katrina on the New Orleans area in 2005

Economic impacts	Social impacts	Environmental impacts
30 offshore oil platforms damaged or destroyed and 9 refineries shut down. This reduced oil production by 25% for 6 months	Over 1 million people evacuated, displaced or made homeless	Storm surge destroyed sections of the barrier islands and Gulf beaches
Forestry, port trade and grain handling severely affected	Most major roads into and out of the city damaged as bridges collapsed	20% of wetland lost, affecting breeding of pelicans, turtles and fish. 16 National Wildlife Refuges damaged
Hundreds of thousands of residents left unemployed. Trickle-down effect with fewer taxes being paid	Overhead power lines brought down by strong winds. Water and food supplies contaminated	5,300 km^2 of forest and woodland destroyed
Total economic impact in Louisiana and Mississippi estimated at over $150 billion	Worst hit groups were those with no personal transport, less well off, non-white and vulnerable	Flood waters containing sewage, heavy metals, pesticides and 24.6 million litres of oil pumped into Lake Pontchartrain

Case Study

Long-term severe weather impacts in New South Wales

Australia, the world's driest inhabited continent, experienced its worst drought in 100 years in 2007, with decreasing water supplies and growing water demand. Rivers saw their lowest-ever October flows in 2007 and Sydney's reservoirs were only 40% full. New South Wales produces 40% of the country's farming output — livestock, vines, cereals and fruit — but these only survived by using large-scale water management and 84% of the country's irrigation schemes.

Direct impacts saw reduced crop and livestock production, water shortages, increased fire hazards and damage to wildlife habitats. Indirect impacts included consequences for farmers and businesses, increased food and timber prices, unemployment, reduced tax revenues and migration.

Examiner tip
Check your understanding of a hazard profile by redrawing Figure 3 on p. 11 for droughts.

Extreme impacts — exam review

Key outcome: designing a fieldwork and research activity to investigate the impacts of an extreme weather event

This fieldwork research is closely linked to the subsection 'Increasing risks', and techniques and approaches could overlap. Fieldwork is likely to be focused on flood hazard, as hurricanes and tornadoes cannot be realistically investigated in the UK. Historical or anecdotal evidence of droughts or heatwaves could be obtained.

Flood fieldwork could take the form of a number of possible options for study:

- local river system near an urban area (especially where new housing is proposed or at risk)
- larger river system (to allow detailed primary measurements)
- hardware models to examine the flood potential of different surfaces

These approaches could be complemented with surveys of people, examining perceived flood risk against actual risk from past records. Surveys can also be undertaken to examine the impacts of a heatwave or drought. These use similar techniques (oral histories, media records). Focus on economic, social and environmental outcomes.

Key outcome: describing and justifying the techniques used to collect fieldwork and research data

Fieldwork could include a range of qualitative and quantitative approaches. Choose from:

- Local flood-risk survey. Generate flood-risk maps, recording land use linked to observations of height. Use interviews and questionnaires to describe past flooding events and their management. Look at the relevance of flood-defence works (value for money, effectiveness).
- River bankfull survey (cross-sections and calculations of discharge). Use measurements of standard channel variables plus bankfull measurements linked to flood risk. Land use adjacent to a floodplain could be mapped.
- Prepare hardware models to simulate flood risk on different surfaces (urban, bare field, grass, etc.). Seek out facts of a known storm (initial response times, total flows etc.).
- Research could make use of historical documents containing flood stories (e.g. newspaper extracts and photos). GIS mapping from the Environment Agency could be helpful, as could Google Earth to model terrain. Secondary river discharge and flood data can be obtained from the National Rivers Flow Archive

Key outcome: describing and justifying techniques chosen to present and analyse findings

Presentation and analysis techniques could include:

- Flood risk — the preparation of individual flood-risk maps and annotation of zones particularly at risk. This could be linked to land-use value. Analysis could use spreadsheets to calculate descriptive statistics and open-sourced qualitative data such as interviews.
- River bankfull plots. Create land-use maps and include descriptive analysis of findings, annotations etc.
- Hardware models — comparison and construction of storm hydrographs from contrasting storm plots.

Key outcome: commenting on data and conclusions; evaluating limitations

Summarise patterns, trends and anomalies revealed by analysis of data observed and obtained from other sources. Provide a summary of the fieldwork process (to what extent is land use influenced by possible flood risk?). How do perceptions of flood risk compare to actual flood risk? How reliable are the results and what factors could have influenced the reliability of flood data (antecedent conditions, unreliable equipment, etc)? How could the design process be modified in terms of locations, range of observation techniques, methods of presentation and analysis?

Increasing risks

Why is the risk from severe weather growing?

The increasing risk posed by severe weather is related to where people live and how vulnerable they are. You should focus your work on examples of storms and flood events. Locations such as Shrewsbury, Tewkesbury, York, Carlisle or Boscastle are good case studies, but you may choose to study a local flood site.

The risk from flooding in the UK is greater than from any other natural disaster. Flooding by rivers (35%) and the sea (65% including erosion) threatens:

- 5 million people in 1.9 million properties, worth £214 billion
- 1.4 million hectares of agricultural land, worth £7 billion

The risk from storms and flooding is affected by factors such as:

- the increasing frequency and severity of extreme weather events, which may be linked to climate change
- population growth along river valleys and coastlines — risks are increasing because of greater floodplain and shoreline occupancy, as seen in the Environment Agency flood map of England and Wales or DEFRA's 'making space for water' policy
- poor management of land, especially the relationship between land use and increased flood risk

Examiner tip
Flood risk is increased by factors such as climate change, population growth and changes in land use.

Flood risk and flood return periods

Flooding has **frequency** and **magnitude**. The flood recurrence interval is an estimate of the likelihood of a flood of a certain size: a flood level likely to happen once in 10 years has a 10% chance of happening in any 1 year, though this is not a forecast and such a flood may happen more than once or not at all.

Knowledge check 8
Explain the term 'flood recurrence interval'.

Investigating flood risks

This is a fieldwork and research opportunity to investigate flood risks associated with a small stream or part of a river catchment. Case studies like York, Carlisle, Uckfield and Boscastle are well documented, but using your own experiences could allow a clearer, more detailed analysis of flood risks at the local scale.

The vocabulary of river basin hydrology is important. Terms such as interception, overland flow, infiltration and stream flow need to be remembered and used carefully when explaining why floods happen. When analysing discharge graphs and how rivers respond to storm events, use terms like lag time, peak discharge and flashy hydrographs to allow rapid explanation.

Flood risk fieldwork could include some of the following:

- interviewing residents as witnesses or victims of flooding
- drawing up a land use map along the course of the stream
- recording evidence of management (bedload traps, barriers, channel widening, realignment, tree planting etc.) as well as sites of potential future problems

- measuring and estimating discharge, bankfull discharge and flood levels
- investigating infiltration and runoff rates

Flood risk research suggestions include:
- consulting local newspaper records, perhaps on microfilm in a public library
- consulting a large-scale OS map of the area, which will help you to judge the scale of the stream
- using online sources such as Google Maps or the Environment Agency website with its postcode search
- using Google Earth maps or photos to present results
- looking at geological and land use maps, to provide valuable data about hydrological change
- obtaining data for UK river gauging stations (including average flow)

Events in Carlisle in 2005 and many areas of the UK in summer 2007 are worth investigating.

Examiner tip
Be an expert in one small-scale case study of flooding.

Increasing risks — exam review

Key outcome: designing a fieldwork and research activity to investigate the increasing risk of an extreme weather event

This is closely linked to 'Extreme impacts' and there could be an overlap of techniques and approaches. Fieldwork is likely to be focused on flooding and linked to other fieldwork carried out to look at impacts. The focus could be on historic or anecdotal evidence of flooding or perceptions of increasing risks from drought or a heatwave.

Flood fieldwork and research could involve two possible contexts/options for study:
- small local river system
- flood risk in a larger river system (York, Boscastle, etc.)

The investigation will require qualitative and quantitative evidence (perceived flood risk compared to actual risk from past records).

Key outcome: describing and justifying the techniques used to collect fieldwork and research data

Fieldwork could involve a mix of qualitative and quantitative approaches.
- Small river system: undertake a local flood-risk survey. Create flood-risk maps recording land use linked to height. Use interviews, questionnaires and oral histories to record past flooding events and their management, plus the perception and relevance of flood defence works (value for money and effectiveness).
- River bankfull surveys: requires measurements of standard channel variables and bankfull measurements linked to flood risk. Map land use adjacent to the floodplain. Record evidence of management (e.g. widening) and include local infiltration studies.
- Flood risk: use local newspaper reports, qualitative observations, photos, oral histories, historic maps and satellite imagery.
- Research: a range of historical documents might support local flood stories (newspaper extracts, photos, etc.). Use GIS mapping from the Environment Agency or Google Earth to model terrain. Get secondary river discharge and flood data from the National Rivers Flow Archive.

Knowledge check 9
Suggest three ways to display information about river flow and the effects of flooding.

Key outcome: describing and justifying the techniques chosen to present and analyse findings

Presentation and analysis techniques available include:

- River bankfull plots (cross-sections and calculations of discharge, with mini graphs/data put on a large-scale base map). Create your own land-use maps and compile a descriptive analysis of findings.
- Preparing individual flood-risk maps with annotations of zones that are particularly at risk. This could be linked to land-use value. Analysis could include use of a spreadsheet to calculate descriptive statistics (mode, mean, median) and open-sourced qualitative data such as interviews.

Key outcome: commenting on the data and conclusions; evaluating limitations

Summarise data patterns, trends and anomalies, as revealed by analysis of data types observed by you and obtained from other sources. Provide a summary of the fieldwork process (can flood risk be calculated using bankfull estimations?). How do perceptions of flood risk compare to actual flood risk? How reliable are the results and what factors could have influenced the reliability of flood data (antecedent conditions, unreliable equipment)?

If the design process was to be modified, how could it be changed in terms of locations, range of observation techniques, methods of presentation and analysis?

Managing extreme weather

How can we best manage extreme weather?

This is a core fieldwork and research opportunity to investigate ways of managing and responding to extreme weather events. It considers how some management strategies are more successful than others. Flood protection and hurricane warnings are the main topics considered.

A useful exercise would be to evaluate the success of existing flood management strategies in a small area, suggesting how these might be developed or improved in the future. Secondary information from organisations such as the Environment Agency or insurance companies would give a fuller picture, as would:

- mapping existing flood defences and strategies
- surveying the purpose, age and condition of flood defences
- completing resident and business questionnaires for views on success and future needs
- examining present and planned flood protection and the Environment Agency's early warning systems
- interviewing people responsible for flood defence schemes
- researching options available and why some were adopted and others not
- relating flood frequency and height to completion of defences or schemes

Flood management case studies such as Boscastle, York or Shrewsbury involve local and integrated catchment solutions for flood management, including hard

Examiner tip
Know how to assess the success (or not) of one flood management scheme.

Examiner tip
Build on the earlier case study you chose, this time focusing on flood management (Table 4 may help).

engineering schemes (such as dams) and less aggressive local schemes, which attempt to restore more natural conditions. Recent measures tend to adopt a more holistic or sustainable approach (see Table 4). Copy the table and use your own examples to complete it.

Table 4 Differing approaches to flood management

Method		UK examples	Advantages and disadvantages
Structural (hard engineering)	Flood storage, e.g. dams, washlands and relief channels		
	Channelisation, realignment and flood barriers		
	Flood protection, e.g. flood proofing and embankments		
Non-structural	Emergency action, warnings and flood relief		
	Flood insurance		
	Floodplain zoning		
Sustainable ('soft')	Whole catchment approach, e.g. planting woodland		
	River restoration		

Knowledge check 10

Explain briefly these two methods of flood management:

- washlands
- floodplain zoning

Wider research into flood and hurricane management — where warning systems, protection, education, evacuation and insurance are intended to reduce disasters — can produce interesting results:

- Hurricane Mitch showed that despite improved levels of technology its track was not accurately predicted. Its unexpected swing over Honduras was due to El Niño conditions, a factor ignored by forecasters.
  - flood rather than storm damage was the main killer
  - in poor countries, warnings are not always effective
- Hurricane Katrina showed that, despite sophisticated systems and finance in the USA, the threat from flooding was underestimated when the levée system was overwhelmed.
  - mistakes were made in implementing the recovery programme
  - vulnerable people/groups were affected disproportionately

Examiner tip

Predicting any extreme weather event is not easy.

Case Study

Flood management in York

Flood management schemes in and around York (see Figure 4) have been designed to combat upstream issues in the River Ouse catchment and increased risks in the city itself. Around York, development has decreased the amount of natural soak-away land and increased drainage networks. Low bridges and building on land near the river have increased the flood risk.

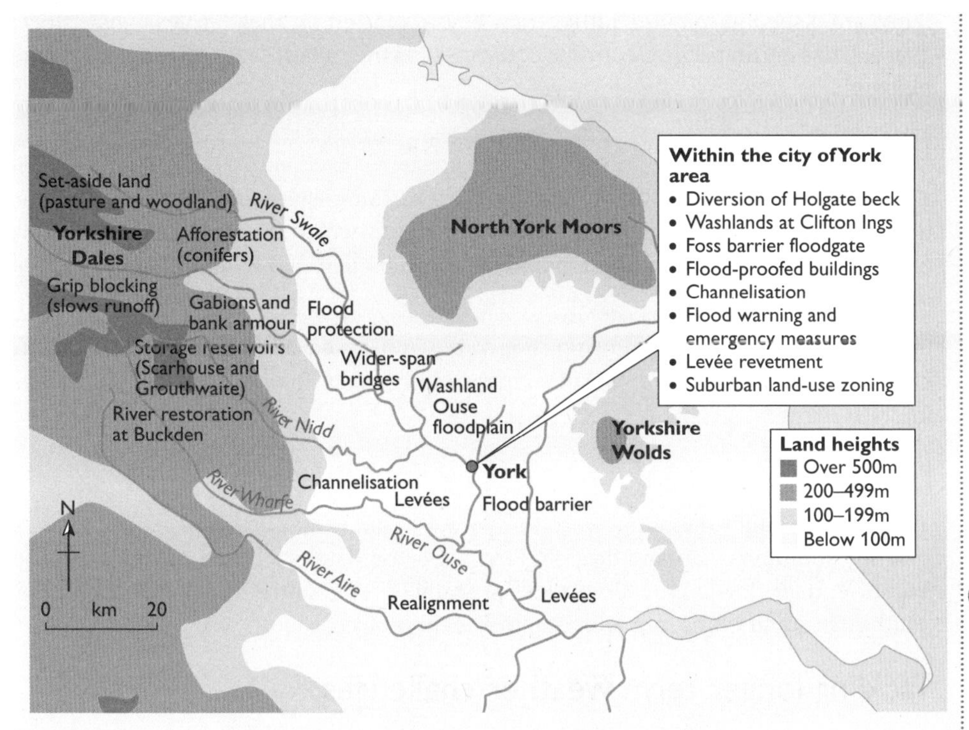

Figure 4 Flood management strategies in and around York

Knowledge check 11

Which of the strategies shown in Figure 4 might be sustainable?

How is new technology helping weather management?

Hurricanes, tornadoes, floods and drought provide examples of how new technology is being applied to monitor and forecast extreme weather events. FEMA (the Federal Emergency Management Agency) and NOAA (the National Oceanic Atmosphere Administration) operate systems in the USA, while the Met Office and the Environment Agency use technology-based systems in the UK.

Technology can also help communities to prepare for extreme weather and reduce weather impact by developing education, safety procedures, communications, engineering, water management and agriculture.

Storms

Information about extreme weather in the USA comes from the National Weather Service and the National Hurricane Center in Miami. These institutions identify, track and report on hurricane activity using weather satellites, ocean buoys and aircraft. Radar detects rainfall patterns and computer models help predict changes in direction, wind speed and the likely landfall.

In the UK the Meteorological Office uses similar technology to observe and forecast weather. Details of weather research, forecasting and severe weather warnings can all be found on the Met Office website.

Tornadoes are researched at the National Severe Storms Lab at Norman, Oklahoma, in the heart of Tornado Alley. Doppler radar enables warnings to be given, though these are issued in minutes rather than days as for hurricanes. Professional 'storm chasers' use cutting edge electronic equipment as they seek to record and understand twisters.

Examiner tip
Improved technology provides our best chance of coping with extreme weather in the future.

Knowledge check 12
Suggest ways in which each of these technologies can help manage weather extremes:
- satellite images
- computer models
- mobile phones

Floods

The Environment Agency monitors river levels and issues flood warnings in the UK, using a network of gauging stations, many of which are linked automatically to flood modelling/simulation computer programs (e.g. the FloodRanger Foresight project).

In the USA river-flow monitoring, flood modelling and forecasting are extremely high-tech. The Geological Service maps daily river flow and uses interactive forecast maps.

Drought

The USA Drought Monitor website (**www.drought.unl.edu/DM/monitor.html**) has a monthly summary map and drought forecasts. The data are based on soil and crop moisture, river flow and reservoir levels. Many other countries, including Australia, South Africa and China, are adopting this technology.

Examiner tip
Technology is increasingly being directed at sustainable rather than hard engineering solutions to extreme weather.

Tackling longer-term weather challenges

This is a big topic, and it is probably best to focus on drought and water management issues. Focusing on southeast England and a contrasting example would allow evaluation of current or future schemes. Other examples could come from Africa, Australia or even China.

Three important ideas to consider are shown in the first column of Table 5.

Table 5 Contrasting solutions to drought

	Southeast England	East Africa
Water collection and distribution	Abstract water from aquifers Take water from reservoirs in Wales Repair leaking infrastructure Government or business decisions	Use bunds, line of stones etc. Fit pumps, repair or dig new wells Community-owned/built facilities Help from aid and NGOs
Adapting farming techniques	Reduce irrigation use Shift to Mediterranean crops Use gene technology	Change from nomads to cultivators Grow drought resistant crops Use of intermediate technology
Recycling and conserving water	Recycle more river water Use more 'grey' water Reduce water footprint (by metering)	Collect and store rain water underground until dry season Separate 'clean' and reusable water

Knowledge check 13
Give an example of each of these methods of drought management:
- intermediate technology
- plant genetics
- water harvesting

Case Study

Contrasting attitudes to drought

Southeast England will see water shortages over the next 15 years. A further 1.1 million households are expected to move to the region by 2020. Thames Water has the worst leakage record of the UK's water companies and is said to be losing nearly 1 billion litres of water each day. The Consumer Council for Water has warned that better water efficiency, new reservoirs and desalination plants will be needed.

Ethiopia is part of the Sahel region of Africa and has long been subject to periodic droughts, most notably in 1984–85, when 8 million people faced starvation and Live Aid raised £5 million in 3 days. Sustainable schemes focus on water conservation and rainwater harvesting rather than larger, expensive approaches.

Managing extreme weather — exam review

Key outcome: designing a fieldwork and research activity to investigate the management of extreme weather

Fieldwork and research for this part of the specification are likely to focus as much on secondary evidence as on primary fieldwork evidence. The design could require contacting those who manage the response and cope with extreme weather. In the UK, the Environment Agency will be the main player for flooding, and the various water companies for drought. Fieldwork could include an evaluation of flood defence schemes (visual impacts etc.).

You could evaluate attempts to reduce extreme weather impacts (drought measures and flood defence schemes through the eyes of residents, businesses, visitors). Compare short with long term, and value for money.

Key outcome: describing and justifying the methods and techniques used to collect fieldwork and research data

Fieldwork can involve a mix of qualitative and quantitative approaches:

- Qualitative field notes, field sketches, photographs of flood defence works/schemes or reservoirs for drought etc. Make use of interviews with Environment Agency staff, oral histories from local residents and focus groups.
- Quantitative structured questionnaires can be used for residents, visitors and businesses looking at positives and negatives.

Draw on information from GIS maps of flood risks and documented evidence such as newspapers to investigate the effectiveness of hurricane warning systems. Historic river flood data from the National Rivers Flow Archive can also be helpful.

Key outcome: describing and justifying the techniques chosen to present and analyse findings

Your choice of presentation techniques will be influenced by data type, but could include a DVD of interviews with accompanying transcript, annotated photos, notes, sketches and a customised GIS flood-risk map. Other data could be processed, tabulated and presented using more traditional approaches such as bar charts. Analysis using simple statistics could also be appropriate (questionnaire data, satellite imagery showing effectiveness of hurricane warning systems, new technology, flow diagrams and mind-maps describing the theories behind flood defence schemes).

Examiner tip
Practise writing up a flood management report stating your results and what conclusions they lead to.

Knowledge check 14
Suggest why data from questionnaires and media sources in your investigations may be unreliable.

Key outcome: commenting on the data and conclusions; evaluating limitations

Summarise data patterns, trends and anomalies as revealed by analysis of data types, both observed by you and obtained from other sources. Describe the overall summary of the fieldwork process (the effectiveness of measures to combat extreme weather — short term compared with longer term). How is technology helping? Compare the various approaches.

How reliable are the results and what factors could have influenced the reliability (sample size/selection of respondents, lack of pre-calibration for environmental quality scores)? How could the design process be modified in terms of locations, range of observation techniques, methods of presentation and analysis?

Summary

Extreme weather

- Extreme weather is severe and unpredictable.
- Know the causes and characteristics of a hurricane.
- Plan how best to investigate the effects of river flooding.
- Understand how the impacts of drought differ from those of other weather hazards.
- Understand how increased flooding has physical and human causes.
- River hydrology provides important clues as to how localised flooding occurs.
- Know how to find useful sources of secondary information on weather and flood management.
- Research the impacts of one small-scale flood event.
- Evaluate a local or small-scale flood management scheme.
- Technology plays a key role in hurricane forecasting and preparedness.
- Reflecting on the results of your fieldwork and research is an important part of any investigation.

Crowded coasts

Competition for coasts

What factors shape the coastline?

This section considers how natural factors create recognisable coastal features, opportunities for development and pressures that threaten valuable environments. Two useful strategies are:

- obtaining resources to illustrate the world's varied coastal scenery — try to cover coastal geology, geomorphology, climate and ecosystems

- focusing on one section of coastline and identifying the natural features found there — summarise the opportunities and challenges they present

Here is a list of geographical resources and features that you could look out for:

- how geology creates attractive coastal scenery, e.g. Dorset's Jurassic Coast World Heritage Site
- upland coastlines providing sheltered inlets and natural harbours, e.g. aerial photographs of Hong Kong or Sydney
- maps of estuaries and their hinterlands allowing easy transport, e.g. Rotterdam or Shanghai
- the value of coastal ecosystems, e.g. Caribbean mangroves and reefs
- beaches and attractive climates, offering great potential for tourism, e.g. a travel brochure for holidays to Florida
- river floodplains, deltas and inshore areas, which provide food resources (farming and fishing) to support large urban growth, e.g. satellite photographs of Cairo and the Nile delta

Examiner tip

Identify the natural features of one small stretch of coastline.

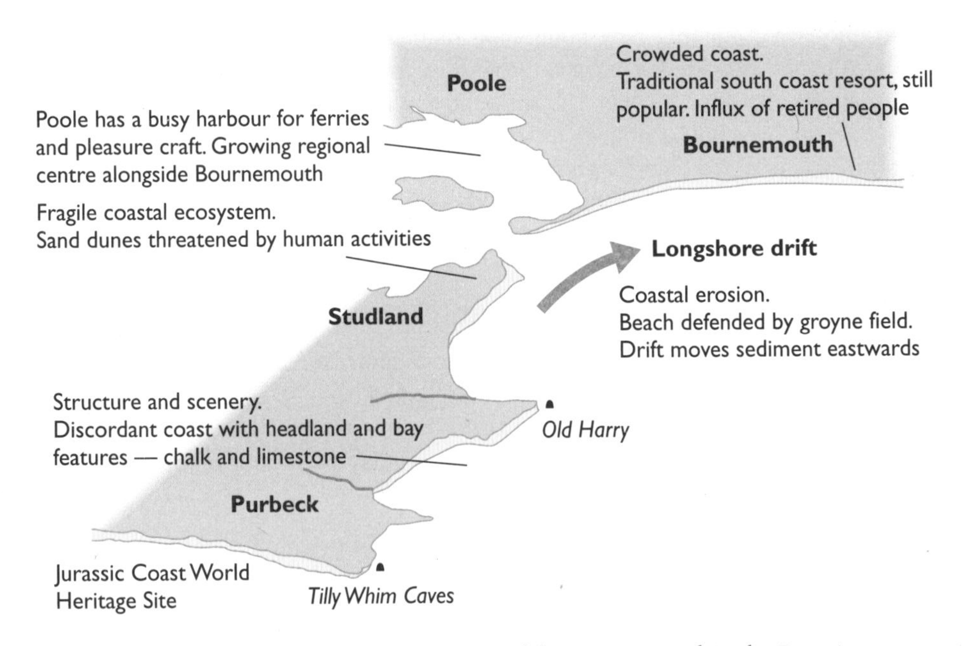

Figure 5 A student's sketch map of the coast around Poole, Dorset

Coastal development brings opportunities and challenges for managers of coastal areas. Conservation of areas with fragile ecosystems is becoming increasingly difficult. The economic benefits of industry and tourism can be counterbalanced by environmental costs such as pollution.

Why is the coastal zone attractive to developers?

Coasts have traditionally attracted settlement. European countries built great ports to receive goods from their colonies abroad. In modern times peripheral populations continue to increase worldwide. Australia has an obvious coastal distribution of population and in the USA, 75% of the population live within 100 km of the sea or the Great Lakes. Recent trends in China show rapid **coastal**

migration. In developing countries these changes are fuelled by new employment opportunities, whereas in Western economies retirement and tourism are the major drivers. Population growth in coastal areas also includes some of the world's major cities.

The following list gives reasons why coasts have encountered such rapid population growth: flat land, soil fertility, equable climate, biodiversity, potential for fishing, recreation/tourism and industrial/port development, and accessibility.

Case Study

Benidorm's rapid growth

Benidorm, on Spain's Costa Blanca, is a major destination for European holidaymakers. It attracts over 4 million visitors annually. Until 1950 it was a quiet village (population 2,726) with a local economy based on tuna fishing and wine production. Changes came about in the early 1950s as fish catches declined and disease damaged the vines. The solution was a package holiday industry based on sea, sun and sand, building new hotels and using foreign tour operators. A new airport at Alicante triggered a massive growth in construction, and the population rose through the 1970s to 25,000. The resident population today is close to 70,000 but this rises to over half a million in the summer. To stay ahead of competition and retain visitors to its blue-flag beaches Benidorm has developed theme parks, nightlife and water sports. Some holidaymakers no longer visit this crowded coast, choosing new destinations and different types of holiday.

Investigating coastal development

This fieldwork and research activity looks at how contrasting crowded coasts have developed over time. Primary data should be collected from an accessible UK resort, and research or a case study should be used to investigate a foreign destination. Research abroad could make use of travel and holiday information, tourist guides and online resources. Florida would be a good choice.

Examiner tip

Investigate the growth and economic development of one coastal resort.

Fieldwork in traditional UK resorts such as Blackpool or Brighton could include:

- classifying and mapping tourist attractions and facilities
- carrying out an age-of-building survey to show the spatial pattern of the resort's growth
- completing a visitor survey using interviews or questionnaires — this is a serious undertaking, so check if data already exist
- carrying out pedestrian, traffic-flow and car-parking surveys during peak holiday times and out of season
- using a transect from the shore inland to reveal changing patterns
- using large-scale Ordnance Survey maps to make fieldwork base maps
- using digital cameras or mobile phones to capture images

Knowledge check 15

Suggest (a) why coastal locations continue to have growing populations and (b) what evidence might show whether a resort is in growth (or decline).

A range of secondary resources to complement this work could include:

- changing population census data or records from local publications
- old maps of resort growth (**www.old-maps.co.uk**)
- images from Google Maps
- historic photographs, e.g. postcards
- local newspaper archives recording past events and developments

Competition for coasts — exam review

Key outcome: designing a fieldwork and research activity to investigate how crowded coasts have developed over time

Design of fieldwork and research is likely to focus on two contrasting localities, although one, a UK resort, will be the main source of primary data. It could be a resort versus a rural beach. Location could be determined by accessibility and historical factors. Compare results in season and out of season, such as weather.

Surveys could use grid patterns to demonstrate land use, a series of radiating transects (systematic intervals), or a transect from the shore to inland. A similar approach could be taken for pedestrian counts in small zoned areas. A stratified sampling approach might be suitable for questionnaires to ensure fair testing.

Key outcome: describing and justifying the methods and techniques used to collect fieldwork and research data

Fieldwork could involve a choice of qualitative and quantitative data. Qualitative data could be oral histories to explore growth, extended interviews, visual evidence of change through DVD/video and digital images, plus annotation of historic maps. Quantitative data could be questionnaires (residents, visitors, businesses), land use maps, building density, pedestrian movement, car parking surveys and building age surveys (patterns of growth).

Research could investigate changing population data from historical records or local publications. Old maps could show growth (say from 1870), and use images from GIS/Google Earth and local newspaper archives.

Key outcome: describing and justifying the techniques chosen to present and analyse findings

The choice of presentation techniques will be influenced by data type — e.g. DVD of transect of change, annotated photos, notes, sketches, images on customised GIS land-use change maps or historical maps. Other data could be processed, tabulated and presented using more traditional approaches (questionnaires, bar charts, gain/loss for bi-polar indices, transect data of building density).

Your analysis could use simple statistics (the mode, mean and median of questionnaire data). Annotated satellite imagery could show patterns of growth and impacts, and flow diagrams and mind-maps could describe the theory of growth in particular areas.

Examiner tip
Use a variety of maps to show the changing land use over time in a coastal resort (e.g. compare Goad maps over a number of years).

Key outcome: commenting on the data and conclusions; evaluating limitations

You should summarise data patterns, trends and anomalies as revealed by analysis of data types observed and obtained from other sources. Provide a summary of the fieldwork process (where are pockets of growth and why are they attractive for development?). How reliable are the results and what factors could have influenced

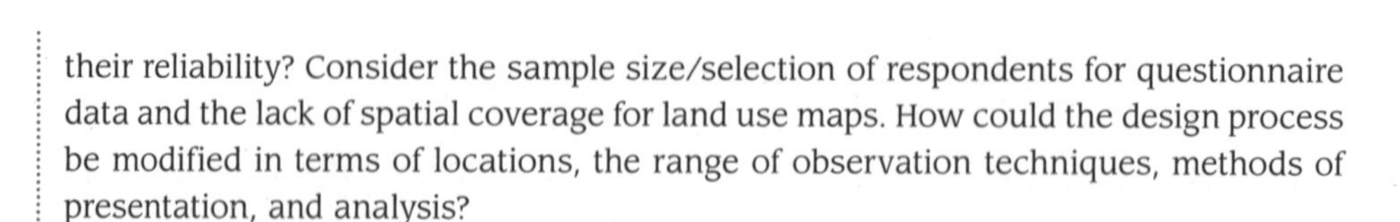

their reliability? Consider the sample size/selection of respondents for questionnaire data and the lack of spatial coverage for land use maps. How could the design process be modified in terms of locations, the range of observation techniques, methods of presentation, and analysis?

Coping with the pressure

How does coastal development increase demands on space?

Coastal developments create patterns resulting from the competition for space. This can lead to pressure on coastal environments, and many resorts and ports have distinctive patterns of land use and buildings. The shoreline can distort the typical patterns of urban land use (often semicircular):

- Hotels and guesthouses demand sea views or promenade sites.
- Tourist/recreational attractions occupy key sites. Blackpool, for example, has piers, a pleasure beach, a golden mile of promenade, tram transport and a famous tower.

This pattern could be disturbed by development and conflicts over land:

- Retail and industrial parks may spoil views and disrupt holiday traffic.
- Some industries or port facilities could clash with tourist activities or fragile environments.
- Major schemes (theme parks, conference venues, marinas, casinos) become big issues in the community.
- Local social, economic and environmental issues will affect a resort's image.

There is a need for effective planning controls as the amount of available land decreases and the rate of development intensifies.

Investigating the impacts of coastal development

This is a core fieldwork and research opportunity, with development and conservation meeting head-on.

The work should cover two aspects:

- the over-use of resources, pollution, and tourist/urban development in beach areas
- impacts seen in fragile, high value, coastal ecosystems such as sand dunes, saltmarshes or vulnerable coastal zones

The first aspect can be tackled by investigating an urban resort while the second can probe threats to a coastal ecological site. Surveys of the recreational and biodiversity value of sites could be linked to how human activity is altering or threatening their survival. Examples could include areas such as Studland, and could assess the impacts of pollution, trampling, litter, grazing and other activities.

A beach pollution survey could include:

- accessibility for the general public
- areas of high pedestrian traffic
- analysis of litter over 25 m of beach.

The websites of the Blue Flag (**www.blueflag.org**) and Quality Coast Awards will allow you to compare fieldwork results with beaches in other coastal resorts.

Examiner tip

Be sure to focus on conflicts and issues that arise in coastal locations.

Knowledge check 16

List four different reasons why a Mediterranean resort could have become more successful than a traditional British one.

An ecological site survey could be used in a threatened sand dune or similar ecosystem, perhaps surveying and analysing differences between a pristine area and a damaged site:

- recording endemic species (plants and animals found only in a particular area)
- using a **transect** to sample the changing **biodiversity** of the area (**zonation**)
- scoring the site's aesthetic, amenity and vulnerability values
- identifying how the area is being threatened by human activities
- evaluating the degree of environmental damage
- discovering what protection, if any, is afforded the site
- interviewing stakeholders for their opinions on causes, impacts and solutions

Knowledge check 17

Describe what each of these could tell you about the impacts of development:

- a beach pollution survey
- a transect through a sand dune area
- a stakeholder questionnaire

Case Study

Ainsdale Dunes, a threatened coastal environment

These dunes on Merseyside are important for wildlife, conservation and scientific purposes. They are the habitat of rare sand lizards and the dune slacks are home to natterjack toads. Natural England looks after the reserve within the Sefton Coastal Zone Management Scheme. Threats to the 7 km^2 of dunes come from natural and human sources:

- non-native species such as grasses, poplar and pine are invading the landward edge of the area
- visitors are responsible for trampling, habitat loss, litter and fires — around 5 million people live within 1 hour's drive
- there are competing land uses surrounding the dunes, including golf courses, forestry and recreation areas

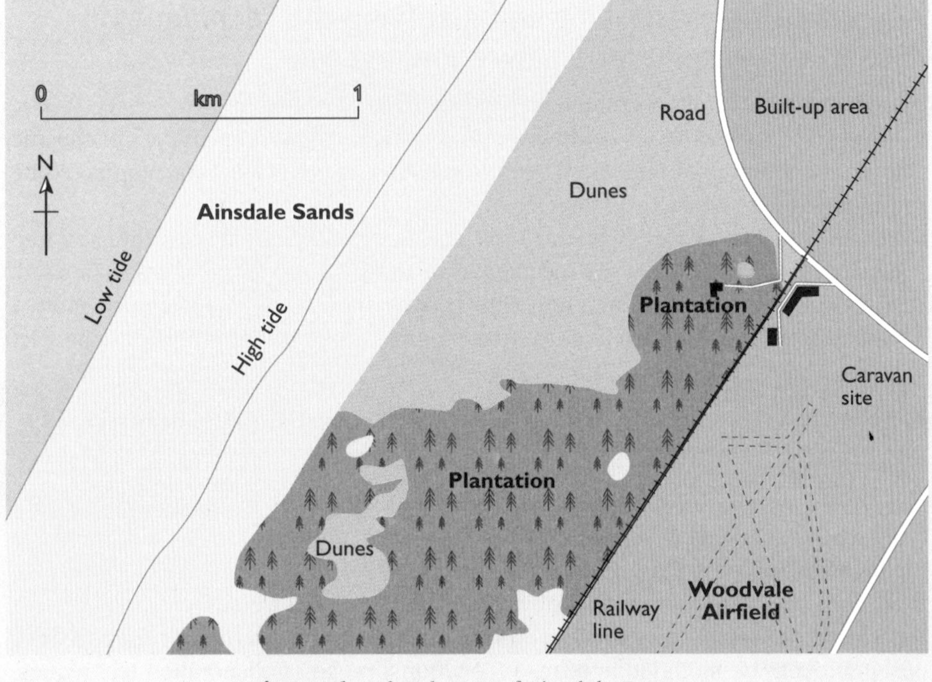

Figure 6 A sketch map of Ainsdale Dunes

Coping with the pressure — exam review

Key outcome: designing a fieldwork and research activity to investigate the impacts of coastal development

Fieldwork in coastal areas can focus on a range of inter-linked themes, such as beach pollution, trampling, litter, visitor surveys and activity patterns. A busy resort beach could be compared with a rural beach that has high ecological value and is relatively undisturbed by people.

Your fieldwork could involve systematic or stratified sampling (frequency and type of litter/plants). In busy areas use questionnaires, interviews and maps showing activity patterns as well as personalised landscape or environmental quality sheets. Interviews with stakeholders and local community groups would be beneficial.

Key outcome: describing and justifying the techniques used to collect fieldwork and research data

Fieldwork could be a mixture of qualitative and quantitative approaches. Qualitative methods could include field sketches, video/DVD, activity maps, focus groups and extended interviews with community groups, resort managers and local authorities. Quantitative methods could be footpath analysis, litter surveys, biodiversity surveys (using plant keys) or an assessment of ecological value. Research could include historical documents — newspaper extracts, postcards, local reports — plus GIS mapping using Google Earth to provide digitised backdrops.

Key outcome: describing and justifying the techniques chosen to present and analyse findings

Presentation and analysis techniques include:

- Transects/cross-sections could be plotted using kite diagrams or bars of specific diversity. Thumbnail images of plants/litter on a large-scale base map could be annotated with causes and degrees of impact.
- Simple statistics using Spearman rank could get R_S value, species diversity and distance from shore etc. Mode, mean, median and Mann-Whitney could also be used.
- Litter surveys can be treated in a similar way, and activity maps can be plotted with different thickness of lines to show patterns and time of movement of people.

Key outcome: commenting on the data and conclusions; evaluating limitations

You should summarise data patterns, trends and anomalies by the analysis of data types, both observed by you and obtained from other sources. Provide a summary of the fieldwork process (how serious are the pressures on the coastal environment?). How does the impact vary spatially (resort versus rural versus protected)? How reliable are the results and what factors could have influenced them? How could the design process be modified in terms of locations, range of observation techniques, methods of presentation and analysis?

Benefits and costs of coastal development

This should be approached using research or a single case study, but could also involve some fieldwork. Locations such as Dibden Bay, Cardiff Bay, Akamas (Cyprus) and Breton Bay (Australia) are examples used in previous examination papers.

Some of the key tools used by planners and other decision-makers are:

- **environmental impact analysis** — considering the likely effects of development
- **cost–benefit** ratio (benefits divided by costs); this is major decision-making tool
- **risk assessment** — looking at the risks of the scheme, e.g. the likelihood of floods or storms
- examining the views and objectives of the main stakeholders
- choosing between options
- analysing impacts of decisions

An issues evaluation exercise could be researched or set up as a desktop study, and a mock public enquiry could explore issues more fully and help prepare for Unit 3 of the A2 course.

Examiner tip

A detailed investigation into the costs and benefits of a development provides a useful link to Unit 3 in the A2 course.

Knowledge check 18

Explain what each of these terms means:

- an environmental impact analysis
- a cost–benefit ratio
- a risk assessment

Case Study

Severn Barrage: clean energy versus the environment

This scheme is a classic study in economic gain versus environmental loss. References provided should help you gather relevant information. Use the headings listed below:

- building costs
- natural advantages of location
- strategy and design of barrage
- economic benefits (power)
- environmental impacts
- views of stakeholders

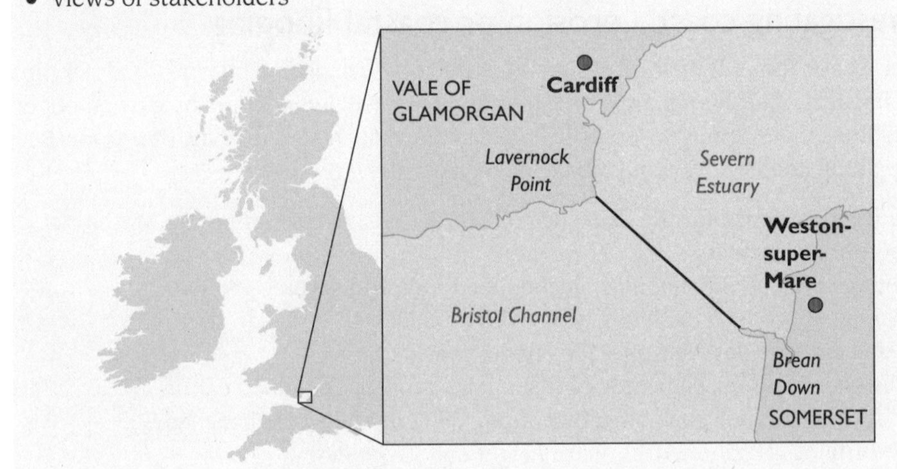

Figure 7 The proposed Severn barrage

Increasing risks

Risks from coastal erosion and flooding

You should be aware of the risks posed by the growing incidence of coastal hazards, and their social, economic and environmental impacts. Increasing the coastal population further increases this risk. There are three different situations to bear in mind:

- rapid erosion — coastlines undergoing significant erosion due to natural vulnerability (weak geology, e.g. Holderness, and/or low relief, e.g. USA east coast)
- increased flooding — coastlines and estuaries at risk from rising sea levels (e.g. Thames Estuary and hurricane impacts in the US Gulf States)
- tidal waves — coastlines damaged by tsunamis, e.g. Aceh (Indonesia), Hawaii and Japan

A useful exercise might be to construct a simple table to compare these situations. Copy and complete the table below and write brief bullets and basic facts. One row has been completed already.

Table 6 A summary of the three coastal risks

Situation and causes/ background	Risk assessment	Range of impacts
Holderness coast		
US Gulf Coast		
Japan earthquake and tsunami, 2011 (9.0 magnitude earthquake causing tidal waves up to 40 m high)	Destroyed infrastructure and settlements on the southern coast of Tohoku, Japan	25,000+ deaths $300 billion cost Damage to coastal settlements and nuclear power plant

Investigating coastal erosion or coastal flooding

This section looks at the rate of coastal retreat or the degree of coastal flood risk, and the impacts of both on people and environments. Fieldwork could be carried out on coastlines that are not necessarily crowded. Sites such as Start Bay, Barton-on-Sea, Mappleton and Porlock would make good examples.

Fieldwork evidence for the existence of and the rate of coastal retreat, or the risk of flooding, can include:

- the remains of old buildings, abandoned roads and severed footpaths
- damage to coastal defences, properties or facilities
- flood damage (lagoons, previous flood levels)
- cliff-face features, such as recent slumping, gullies and fallen debris
- cliff-foot features, such as undercutting, cliff fall and storm beaches
- identifying longshore drift, wave height and frequency

Knowledge check 19

List (a) four different pieces of fieldwork evidence that may suggest a coastline is eroding and (b) four sources of information that can prove long-term coastal erosion.

Questionnaires and interviews with local residents, businesses and officials could reveal social and economic impacts.

Local research offers a wide range of evidence:

- Environment Agency data of risk and rates of erosion, and estimates of flood and storm return periods

- universities often monitor changes in great detail
- council planning departments will have records of changing land use
- documents, photographs and old maps show coastline changes
- satellite images might show wave and drift patterns
- environmental losses might be recorded by conservationists
- local newspapers provide a rich source of information about how local people are affected by rapid erosion or flooding events
- present-day erosion data and maps of reconstructed coastlines can enable you to work out an average rate of coastal retreat

Examiner tip

Practise drawing a sketch map of a stretch of coastline (e.g. Figure 8), plotting evidence of erosion or flooding.

Case Study

Coastal retreat at Holderness

Erosion on this part of the UK's east coast is caused by storm waves, longshore drift and soft glacial clays. It is one of the fastest eroding coastlines in Europe (1.2 m per year on average). Climate change (in the form of more severe storms) is accelerating this cliff erosion, and the low-lying land around the River Humber estuary is increasingly at risk from sea flooding.

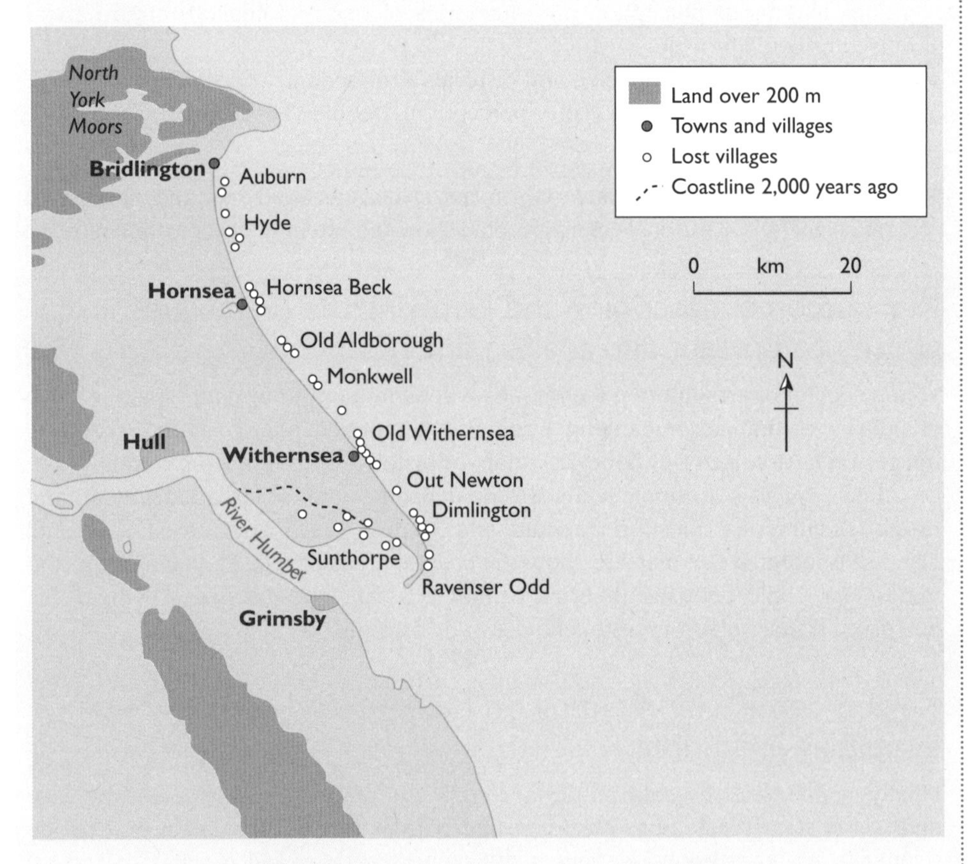

Figure 8 The lost villages of east Yorkshire

Increasing risks — exam review

Key outcome: designing a fieldwork and research activity to investigate risks, coastal retreat or floods

Fieldwork and research for this part of the specification will focus on either rates of coastal retreat or investigating coastal flood risk. The type of location will determine your choice of fieldwork. Both options could be covered if a stretch of coast is vulnerable to both flooding and coastal retreat. Note that the second option must be coastal flooding and not river flooding.

Key outcome: describing and justifying the methods and techniques used to collect fieldwork and research data

Fieldwork could involve a range of possibilities:

- Rates of coastal retreat could consider cliff erosion surveys (scales of risk and micro-scale maps of cliff stability and physical features). Annotated photographs and DVD/video could form an evidence base, alongside interviews with focus groups looking at perception and memories of risk and coastal retreat. Local museums could provide evidence.
- Coastal flood risk might involve oral histories and accounts of past events through interviews (residents versus visitor perception). Detailed base mapping could use GPS maps of areas at risk.

Supporting research could include GIS maps of coastal flood risk and newspaper reports of cliff falls. Historical OS maps could show the extent of old coastline retreat.

Key outcome: describing and justifying the techniques chosen to present and analyse findings

Your selection of presentation techniques will be influenced by data type (e.g. DVD of interviews and accompanying transcript, annotated photos, notes, sketches, images on customised GIS flood-risk map, or historical map showing retreat of the coastline). Analysis of simple statistics could also be appropriate (mode, mean and median of numerical questionnaire data and the positives and negatives). You could also use annotated GIS maps to show the coastal areas at risk. Flow diagrams and mind-maps could describe theories behind coastal flood-defence schemes. Use processed transcripts from a stakeholder or decision-maker.

Key outcome: commenting on the data and conclusions; evaluating limitations

Finally, summarise the data patterns, trends and anomalies as revealed by your analysis of data types, both observed and obtained from other sources. Provide a summary of the fieldwork process. What are the rates and impacts of coastal erosion, or what are the impacts and risks associated with coastal erosion? Consider the reliability of the results and what factors could have influenced their reliability (e.g. the sample size and selection of respondents for interviews, the location of cliff erosion, and the possibility that the survey might not be representative of the whole coast). How could the design process be modified in terms of locations, range of observation techniques, methods of presentation and analysis?

Coastal management

Coastal management: changing ideas

You should be aware of the full range of coastal defence strategies and techniques and how these relate to what is feasible, cost effective and appropriate (see Table 7). In your work you could study a number of different places or concentrate on a small section of coast that employs a combination of defence measures that operate together.

Examiner tip
Be aware of the wide range of options available for coastal defence.

Hard engineering involves building a physical structure, usually from rocks or concrete, to protect the coast from the forces of nature. It is used to reduce erosion and the risk of flooding.

Knowledge check 20
Explain the differences between hard and soft coastal engineering.

Soft engineering makes use of natural systems, such as beaches or salt marshes, to help with coastal defences. The advantage of soft systems is that they can absorb and adjust to wave and tidal energy and have a more natural appearance.

Table 7 Coastal engineering techniques

Technique, nature and purpose	Strengths	Weaknesses
Breakwaters — deflect and reduce the power of waves	Can be built from waste materials to mimic the protective nature of reefs	Deflected waves could damage foundations or cause erosion elsewhere. Possible ecological impacts
Embankments — of material such as clay above high tide level to prevent floods in low-lying areas	Simple, relatively cheap and often quite effective	Coastal squeeze reduces protection provided by salt marshes on the landward side
Gabions — small rocks held in metal cages. Could be stacked to build walls	Has some of the strengths of the two techniques above, but costs more	Relatively small-scale solution. Metal cage may fail relatively quickly. If stacked, cages can move during storms
Groynes — wood or rock barriers preventing beach material from being moved by longshore drift	Low capital cost and easily repaired	Likely to interfere with sediment budget, causing deficit or even starvation downdrift
Revetments — sloping ramps placed to absorb the full force of waves. Made of rocks, concrete or timber	Cheaper to construct than sea walls and less at risk to undermining	Do not cope well with powerful storm waves. Could damage foreshore ecosystems
Rip-rap — large rocks placed at foot of sea walls or cliffs to absorb wave energy	Effective, cheap and prevents undermining. Can look relatively natural	Could shift in very heavy storm conditions or be under-scoured by backwash

Examiner tip
Be an expert in a small number of engineering techniques (see Table 7).

Knowledge check 21
Evaluate the comparative successes of groynes and sea walls.

Technique, nature and purpose	Strengths	Weaknesses
Sea walls — to reflect rather than absorb wave energy, and tackle erosion and flooding	Reasonably effective. Used to protect valuable or high-risk property	Costly to build and maintain. Foundations can be undermined on beaches or where there is strong longshore drift
Cliff drainage — removal of water from rock strata by pipes to reduce landslide risk	Cost-effective	Drains can weaken the cliff. Does nothing to prevent rock falls from dry cliffs
Cliff fixing — iron or steel bars used to stabilise cliff face and absorb wave energy	Simple and reasonably cheap	Only suitable for some types of rock. Does not prevent wave erosion
Cliff regrading — the cliff angle is lowered to reduce chances of collapse	Works well on clay cliffs	Effect is to retreat cliff line and use up large area of land
Beach nourishment — sand pumped from seabed to replace any eroded from the beach	Natural-looking process	Expensive and entails long-term commitment. Adverse ecological impacts
Do nothing	Allows time for new research, etc.	Unpopular locally. Inaction could lead to problems at a later date
Dune regeneration — structures and planting to reduce wind speeds and increase sand deposits	Generally effective if managed properly	Only succeeds if public access strictly controlled
Managed retreats — no new developments. Compensate owners for loss of land/property	Sensible long-term strategy. Cost-effective in the long term	Difficult to persuade people they are safe, or that the authorities care
Red-lining — coastal retreat permitted to agreed new coastline inland	Cost-effective	Unpopular at local level, particularly among those losing land or property

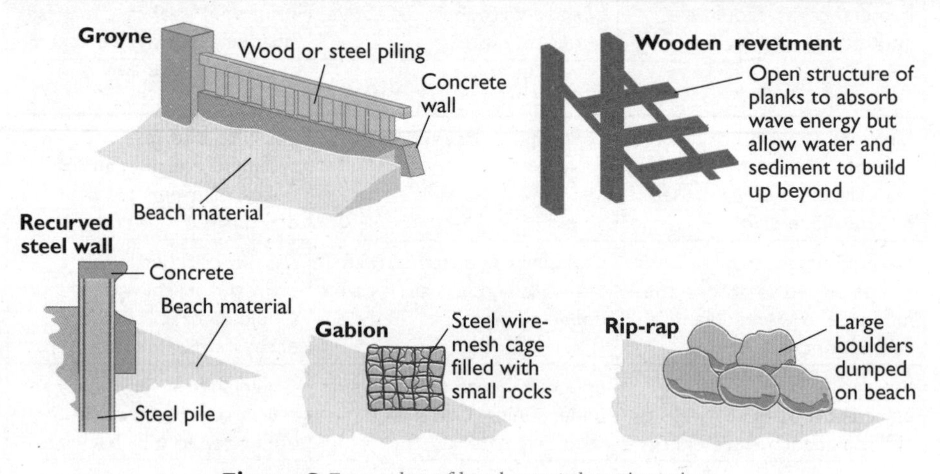

Figure 9 Examples of hard coastal engineering

Examiner tip

Sketch examples of different types of hard defences (e.g. Figure 9) adding labels to show how they work.

Knowledge check 22

Explain (a) beach nourishment and (b) managed retreat.

Investigating coastal management

This part of the specification looks at the success of coastal management strategies. You will need to investigate the following two aspects of coastal management:

- the success of defence schemes along a short stretch of coastline
- the value of strategies designed to manage a coast of high environmental quality

Judging the success of defence schemes

Existing coastal defence schemes could provide useful sites for fieldwork and research. Some examples of inquiry questions are set out below, and they can be adapted to specific locations.

Table 8 Analysing the success of coastal defence schemes using fieldwork and research

Question	Primary data/fieldwork	Additional research
Which techniques are being used?	Identify and plot seawalls, groynes, beach nourishment, etc.	Consult published plans and engineering reports.
Are the choices appropriate?	Does solution match cause? For example, groynes used if longshore drift evident	Engineering report, geology map, feasibility study
What state are the defences in?	Seawall foundations could be undermined by scour	
Are the techniques cost-effective?		Official cost–benefit analysis, value of schemes/components
Are there any unforeseen impacts?	Knock-on effects of longshore drift along the coast, cliff slumping	Maintenance costs report
Are there ecological implications?	Impact on dunes or marshes, pollution affecting fishing etc.	Compulsory environmental impact analysis, conservation reports
Are the stakeholders happy?	Survey (questionnaire, interviews) of residents, businesses etc.	Local newspaper coverage. Environment Agency views
Are there notable improvements?		Research data from scientific groups, university departments

Examiner tip

Be able to evaluate the success of a coastal defence scheme.

Construction costs are important and decision-makers should be aware of the relative values of different defence systems (see Table 9) and must also take into account additional expenditure on maintenance, land purchase, insurance and compensation.

Table 9 Typical construction costs of coastal engineering

Protection technique	Approximate costing	Typical lifespan (years)
Concrete sea walls	£5,000–10,000 per metre	50–75
Earth embankments	£2,000–4,000 per metre	Variable
Revetments	£2,000–4,000 per metre	Less than 50
Groynes	£20,000 each	25–40
Gabions	£1,000 per metre	10–30
Rip-rap	£1,000 each rock	Short

Knowledge check 23

Explain the apparent link between the costs and lifespan of coastal defences shown in Table 9.

Assessing management schemes in high quality environments

You could choose to investigate a honeypot, heritage or similar site, such as the Jurassic Coast World Heritage Site, but the most obvious choice is to look at the management of a coastal ecological site. Sand dunes and salt marshes could be used to evaluate management strategies. Consider comparing data from pristine and damaged areas of dunes. Here, 'damaged' versus 'managed' would be useful. The results of this investigation could be shown as in the following Ainsdale Dunes case study.

Case Study

Management of the Ainsdale Dunes Reserve

The main threats to the reserve are overuse by the public, invasion by rank species and competition from surrounding land use. The methods used and their degree of success have two aspects.

Ecosystem management:

- Scrub cutting using bulldozers to remove plants and tree stumps. This is effective but also expensive and best used as a last resort.
- Flail mowers control the spread of creeping willow, and cut material can be used to surface paths or make compost. There are concerns about timing and ground compaction.
- Scraping involves excavating the dune slacks to increase access to water in drought conditions. This has helped natterjack toad breeding and encouraged the rare petalwort. New slacks should develop naturally.
- Grazing is important to control overgrowing. Rabbits have done this naturally but selected access for Herdwick sheep and some beef cattle is proving extremely successful.
- Sand management to create open areas is important to the succession and for species such as lizards. The threat of erosion and blowouts makes this a risky strategy in some areas.

Visitor management:

- This can involve a zoned system of closed (sanctuary) areas, permit sections, and sites open to the public (encouraging people to stay on the beaches).
- Facilities provided to help and educate include car parking, boardwalks, fencing-off, waymarked trails, information boards and a warden service.

Management strategies for the future

This final item looks at more sustainable and integrated approaches to coastal management that have only recently been adopted in the UK. The two most discussed strategies are coastal realignment and shoreline management planning. These try to accommodate, copy or work alongside natural systems and processes. Ecosystems often play a key role. Their main advantages are:

- they are environmentally friendly and may offer a longer term solution
- they are less expensive than existing defences (economically and environmentally)
- they can include the impacts of global warming and coastal squeeze

Examiner tip

Be an expert in one small case study of successful ecosystem management (e.g. sand dunes).

Coastal realignment

This controversial measure involves allowing the sea to cover previously reclaimed land. It is already happening along the Essex (Blackwater estuary), Kent and

Humberside coastlines. Salt marsh ecosystems are often involved as they can trap sediment and create new land and habitats. This deliberate flooding helps wildlife and saves expenditure on failing seawalls. Figure 10 shows how this has come about, with stage 4 illustrating the managed retreat.

Figure 10 Coastal realignment (managed retreat) along North Sea coasts

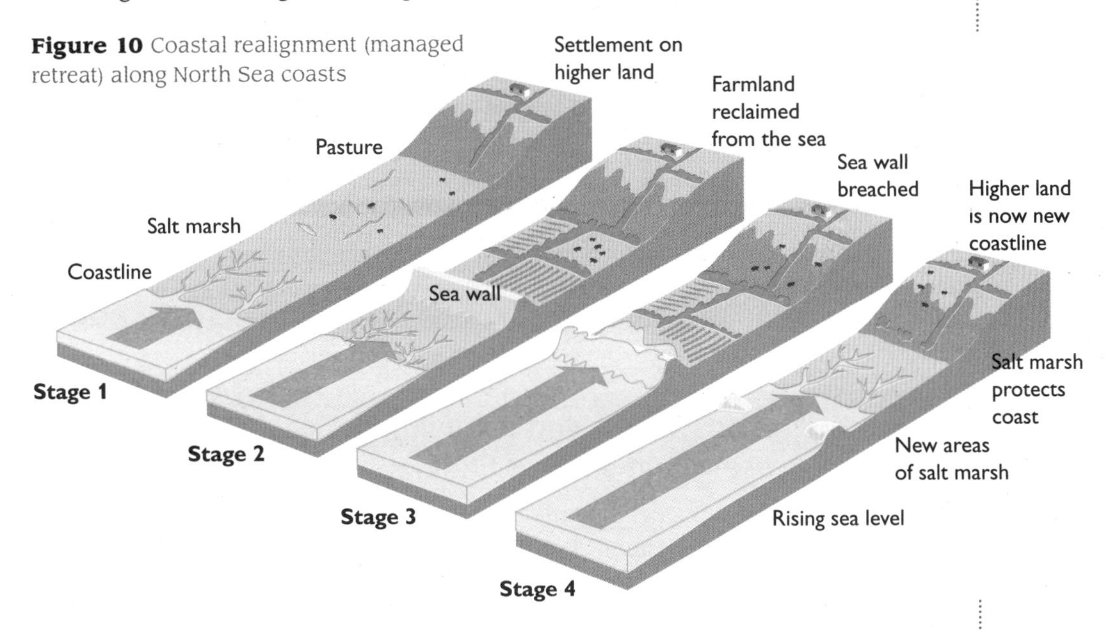

Shoreline management plans (SMPs)

This strategy uses a combination of systems to provide the best long-term solution to defence issues. Hard engineering could protect important economic activity, with softer options being used for a beach or valued ecosystem. The overall package is likely to be sustainable — economically, socially and environmentally. Most UK coasts now have an SMP in place, and you should explore any such locations that are close to where you live.

Knowledge check 24

Explain briefly these terms used in coastal management:

- coastal realignment
- SMPs

Integrated coastal zone management plans (ICZMs)

These deal with a broader range of issues, such as transport and business, in regions where the coast plays a significant role.

Coastal management — exam review

Key outcome: designing a fieldwork and research activity to investigate coastal management schemes and options

Fieldwork and research for this part of the specification could focus on two contrasting localities — a high value ecological site versus a site that has suffered or could suffer development pressures. It could be linked to fieldwork on increasing risks. Many areas lend themselves to this kind of study, especially traditional resorts coming under pressure from changes in culture, visitor type or architectural ambience. Fieldwork could also look at the effectiveness of coastal defences. Studies could focus on the ecological impacts of development (loss of biodiversity/individuality due to development or change in land use). This study could be carried out in a sand dune or salt marsh environment.

Key outcome: describing and justifying the methods and techniques used to collect fieldwork and research data

Fieldwork possibilities could include:

- Coastal defences — you could plot the lines of sea walls, carry out surveys of the condition of defences (for environmental quality) and look at longshore drift to test the effectiveness of defences. Use structured questionnaires (stakeholders, residents, visitors and businesses) and look for evidence of slumping and erosion.
- Consider undertaking ecological surveys, trampling surveys and litter surveys.

Examiner tip

Be prepared to write about:

- your fieldwork activities – what you did and the results this gave you
- your research – what sources you used and what they showed

Supporting research could include maps of changing land use, past visitor surveys (including student data and published sources), planning applications and newspaper coverage. Research data relating to management, species diversity and coastal defence schemes could come from scientific groups and universities.

Key outcome: describing and justifying the techniques chosen to present and analyse findings

Presentation will be influenced by the data type (e.g. cross-section and transect data of a sand dune ecosystem). Kite diagrams could be used for species type, diversity and richness. Other data could be processed, tabulated and presented using traditional approaches (questionnaires, bar charts etc.). Analysis of simple statistics could also be appropriate (mode, mean and median or Spearman rank/Mann-Whitney). Annotated satellite imagery could be used to show the effectiveness of coastal defences. Flow diagrams and mind-maps could describe theories behind coastal defence schemes and management practices, and cost/benefit analysis plus processed transcript information could use various coding methods.

Key outcome: commenting on the data and conclusions; evaluating limitations

You will need to summarise data patterns, trends and anomalies as revealed by the analysis of data types, both observed by you and obtained from other sources. Provide a summary of the fieldwork process (how effective are coastal defences or management options, short term versus longer term?). Describe possible conflicts between decision-makers and local residents. How reliable are the results and what factors could have influenced their reliability (e.g. the sample size and selection of respondents for questionnaires, a lack of pre-calibration for environmental quality scores, misidentification of plants etc.)? How could the design process be modified in terms of locations, the range of observation techniques, methods of presentation and analysis?

Summary

Crowded coasts

- Coastlines vary naturally and as a result of economic development.
- Coastal areas attract population and economic growth.
- Plan how best to investigate the growth and development of at least one coastal resort.
- Understand how conflicts can arise from the impacts of development on a coastline.
- Know how to find useful sources of secondary information on coastal management.
- Research details of hard and soft engineering techniques.
- Evaluate the costs and benefits of a local or small-scale coastal defence scheme.
- Understand the role of ecosystems in planning coastal strategies.
- Realise the importance of sustainable coastal management.
- Reflecting on the results of your fieldwork and research is an important part of any investigation.

Unequal spaces

Recognising inequality

What is inequality?

Inequality is about unevenness — the 'haves' and 'have-nots'. **Spatial inequality** exists where the distribution of resources, wealth and opportunities are not evenly spread. **Economic inequality** is linked to the uneven distribution of wealth or money in society. Many services or opportunities are linked to an ability to pay for them, including healthcare, housing and education.

Knowledge check 25

Describe the concept of spatial inequality.

Social inequality is lack of access to housing, healthcare, education, employment and status. **Technological inequality** is the idea that different people have different standards of access to technology (mobile phone reception, fast broadband linkage, affordable computers etc.). **Environmental inequality** concerns people who are both socially and economically disadvantaged and who live in the worst environments. The overall quality of our environment is improving, but this picture can vary widely between different areas and communities. For example, the most deprived parts of England often have poor air quality, less access to green space and inadequate housing.

Examiner tip

Many of the data response questions that form part (a) of an exam question are based on describing patterns of inequality linked to a resource. Make sure you practise your 'describe' and 'comment on' skills.

Two other concepts to consider are **vertical inequality** — the difference between rich and poor — and **horizontal inequality**, where people of similar backgrounds, status and qualifications have differences in income. Horizontal inequality is the theme of this part of the unit.

The processes leading to inequality

The three most significant factors that contribute to inequality are described in Table 10.

Table 10 Factors contributing to inequality

Factor	What it means	Processes
Access to services	Some social groups are disadvantaged with respect to reaching essential goods and services	People could be disadvantaged due to a lack of services in an area (employment, education or health facilities), difficulty in reaching services (lack of transport), and for social reasons (effects of income, age, ethnicity, disability and social class)
Quality of life	The level of social and economic wellbeing experienced by individuals or groups — various indicators make up the concept	Certain processes can create a downward spiral in quality of life, such as poverty and social exclusion, poor housing and health, fear of crime, low community spirit, lack of services, limited business opportunities, and a poor quality local environment
Economic opportunity	Availability of financial resources (assets) linked to labour market and tax system	Processes leading to local variations in economic opportunities include income, health deprivation, employment, crime, education, skills, and barriers to housing or services

Case Study

Gated communities and urban inequality

A **gated community (GC)** is a residential area where walls, fences or landscaping form physical barriers to entry. Access can be restricted to homes and to streets and neighbourhood amenities. GCs are often characteristic of self-governing homeowner associations who manage the area using their own rules. Residential segregation, by use of GCs, is the most devastating form of urban inequality. Recent research in America suggests that gating reinforces segregation and urban disadvantages. A survey of large cities in the USA found a positive connection between the number of gated communities and the degree of segregation.

Examiner tip

Know some examples of rural and urban inequality supported by facts and figures. These often form part (b) (15 marks) or part (c) (10 marks) of the exam question.

Exploring patterns of spatial inequality

This investigation must examine rural and urban areas and use primary surveys and secondary data. The list below suggests ideas for primary survey techniques:

- public transport routes and frequency of services
- maps of accessibility, e.g. wheelchair access in pedestrian areas
- questionnaires — travel time to particular shops, services and functions
- parking restrictions
- zones of exclusion, e.g. night-time facilities
- use of focus groups and interviews to explore personal inequalities

Knowledge check 26

What are the differences between focus groups and interviews?

Recognising inequality — exam review

Key outcome: designing a fieldwork and research activity to explore patterns of inequality in urban and rural areas

Contrasting sites and areas should be selected for an inequality audit (rural, urban, coastal). You should make reference to different types of sampling such as systematic, stratified or random. Discuss how and why areas were divided up (the number of researchers and manpower) and the size of each zone audited. Justify chosen routes along transects or through areas. Use GIS to help in decision-making.

Key outcome: describing and justifying the techniques chosen to present and analyse findings

Presentation and analysis techniques include the following:

- Transects/cross-sections could be plotted using kite diagrams or bars of specific diversity. Thumbnail images of plants/litter on a large-scale base map could be annotated with causes and degrees of impact.
- Simple statistics using Spearman rank could get R_S value, species diversity and distance from shore etc. Mode, mean, median and Mann-Whitney could also be used.
- Litter surveys can be treated in similar way, and activity maps can be plotted with different thickness of lines to show patterns and time of movement of people.

Key outcome: describing and justifying methods and techniques used to collect fieldwork and research data

Fieldwork could involve a range of qualitative and quantitative data linked to assessment approaches. Qualitative data could be field notes, field sketches, photographs/DVDs, interviews, focus groups, and mobility and accessibility maps. Quantitative data could be questionnaires and surveys of shopping quality, footfall counts, pedestrian counts, customised environmental quality assessments, land use, visitor catchments, litter and graffiti. Use the internet to research geo-demographic data (Acorn and Cameo profiles), and socioeconomic profiles from National Statistics. Geo-located pictures could help with identity (Flickr, Panoramio, Geograph etc.).

Key outcome: describing and justifying the techniques chosen to present and analyse findings

Choice of presentation techniques will be influenced by data type. Quantitative data lend themselves to graphs such as line, scatter and histogram. Qualitative analysis could use more descriptive narrative techniques (describing a photograph, for example). Some data can be spatially represented (thumbnail pictures of evidence of inequality on a large-scale base map of the study area). Analysis using simple statistics may also be appropriate (mode, mean and median). Inter-quartile ranges could be used for some of the quantitative data collected, such as shopping quality, footfalls etc. Descriptive analysis of qualitative data should be considered (open-coding, geographical narratives, abbreviating extended interviews), and a written commentary could accompany a video or photographs of examples of inequality.

Key outcome: commenting on the data and conclusions; evaluating limitations

You need to summarise data patterns, trends and anomalies as revealed by an analysis of data (questionnaires, pedestrian flows, densities). Provide a summary of the fieldwork process. To what extent is the area unequal? How reliable are the findings regarding qualitative and quantitative data? (Too much of one may lead to unreliability.) What other factors could have affected results? (Poor weather when taking photos could make the place look less attractive.) How does first-hand fieldwork data compare to internet demographics and images of inequality? How could the design process be modified in terms of sample location, techniques, methods of presentation and analytical tools?

Inequality for whom?

Social and economic exclusion in rural and urban areas

Inequality, social exclusion and polarisation (growing differences between the 'haves' and 'have-nots') are divisive and socially damaging. Inequality in the UK exists in a variety of environments including urban, rural and coastal districts:

- The lowest wealth and highest poverty rates are concentrated in large cities, including industrialised (or former industrialised) centres such as Birmingham, Liverpool, Manchester, Sheffield and Newcastle.
- The highest wealth and lowest poverty tends to be clustered in the southeast of England (with the exception of inner London).

Table 11 compares exclusion and polarisation in urban and rural areas.

Knowledge check 27

Look up a definition or classification of both rural and urban areas.

Table 11 Exclusion and polarisation in urban and rural areas

Urban areas	Rural areas
Main problems: marginalisation, personal income inequalities	**Main problems: household deprivation, opportunity deprivation, mobility inequality**
• Lack of high quality environmental spaces (public parks/green areas) • Fuel poverty • Segregated schooling and lack of influence on school choice • High cost of inner city transport • High rental prices/lack of accommodation (refugees and illegal immigrants) • Clustering and segregation in ghettoised environments = downward spiral	• Lack of affordable housing • Limited access/choice of healthcare • Lack of choice/high price of local retail provision • Limited public transport • Limited employment opportunity • Fuel poverty (limited choice of fuel types) • Lack of access to technology — high cost of alternatives (satellite, broadband)

The spiral of rural decline (Figure 11) is also an important idea.

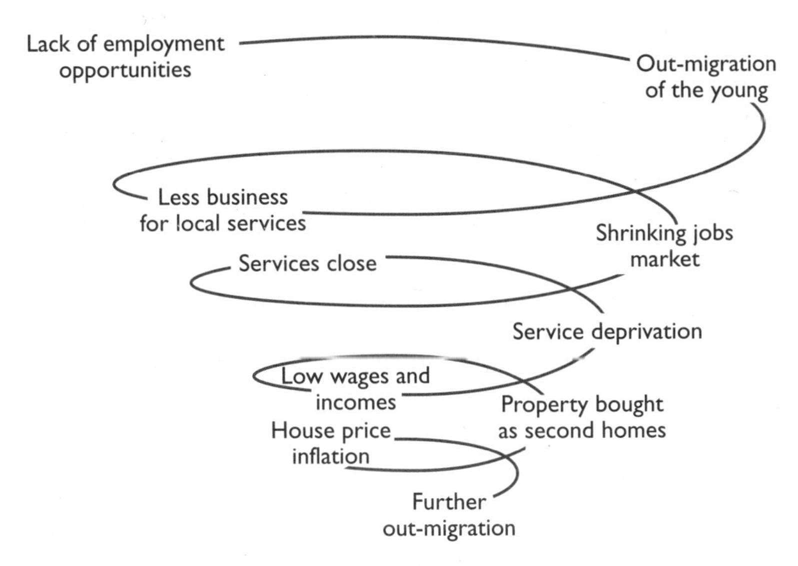

Figure 11 The spiral of rural decline

Examiner tip

Practise drawing a sketch of Figure 11 and getting the labels correct.

Marginalised groups

Marginalisation is linked to exclusion. It can include people or groups who are **economically marginalised** (students, single-parent families, carers), **socially marginalised** (refugees, ethnic minorities, disabled or chronically sick people), or **legally marginalised** (illegal immigrants, the homeless, participants in drug cultures).

Figure 12 Concept diagram for rural deprivation

Case Study

The marginalisation of young people in southwest England

Young people in rural areas often experience difficulties with transport to education, work and social facilities. A recent study of people between the ages of 15 and 24 in southwest England highlighted the following issues:

- High fares and poor publicity about public transport prevented young people from using the limited public services. Young bus passengers felt discouraged by the unfriendly attitudes of bus drivers and older passengers.

- People below driving age rely on lifts from parents for most journeys. Those living in households with one car or no car are at a disadvantage.
- The end of compulsory schooling at 16 presents travel problems. Over 40% of those aged 15–16 say transport issues influence decisions about further education. Limited public transport in rural areas means those entering employment or training are restricted in where and when they work.
- Young people who get financial help (usually from parents) towards driving lessons and buying a car will learn to drive and own cars at a younger age than those who get no help. Parents may continue to help with loans to keep cars roadworthy.

Facts adapted from Joseph Rowntree Foundation Research

Marginalisation is also evident in urban areas, where long-term unemployment can bring serious individual and social problems. The loss of industry (closure of coal mines, steel factories etc.) can be linked to increases in unemployment. The long-term unemployed cannot afford holidays, visits to restaurants, pubs and cinemas. Efforts to save money include missing the dentist to avoid paying fees. Problems can get worse if people go into debt. In all parts of England, those from ethnic minorities are more likely to live in low-income households than white British people. The differences are, however, much more acute in inner London, Manchester, Birmingham, Leicester and Nottingham. More than half of people living in low-income households in London are from ethnic minorities (see Figure 13).

Knowledge check 28

What is the formal definition of 'low-income households' for the UK?

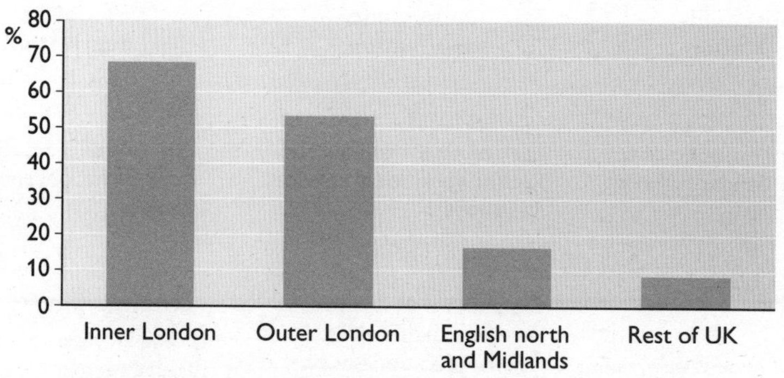

Source: *Households Below Average Income*, DWP; the data are the average for the years 2003/04 to 2005/06; updated February 2008

Figure 13 Proportion of people from ethnic minorities living in low-income households in the UK

Fieldwork and research into inequality

Fieldwork, research and opportunities are linked to the areas of polarisation, exclusion and inequality. Table 12 provides examples of evaluation 'kits' that can be used to investigate various aspects of inequality.

Table 12 Evaluation types for investigating inequality

Survey type	Criteria
General landscape evaluation	Based on gut feelings, i.e. boring/stimulating, ugly/attractive, crowded/peaceful, threatening/welcoming, drab/colourful etc.
Scale of visual pollution	Score 0–3 (no pollution to badly polluted). Criteria might include litter, smells, state of buildings
Likelihood of being burgled	Note presence or absence of burglar alarms, security cameras, metal bars on windows, metal shutters, Neighbourhood Watch stickers etc.
Graffiti assessment	Consider size of words, size of pictures, style of writing, visibility, method of writing/inscription (pen, paint etc.)
Physical condition of buildings and index of decay	Criteria could be deterioration of walls, peeling paint, slipped tiles, broken glass, broken gutters etc.
Shopping survey	Look at shopping opportunities and appearance of streets, including quality and type of shops, other land use, quality of goods, number of vacant shops, street appearance, pedestrian safety, cleanliness etc.

Knowledge check 29

Briefly explain how you might undertake a shopping survey.

Inequality for whom? — exam review

Key outcome: designing a fieldwork and research activity to explore patterns of inequality

There is considerable overlap between 'recognising inequality' and the section 'inequality for whom?' Select contrasting sites and areas for an inequality audit (rural, urban, coastal). The specification involves looking at inequality, and considering various types of evaluation sheets would be sensible. Create an inequality checklist. Systematic, stratified or random sampling could be used in a survey.

Discuss how and why areas were divided up (the number of researchers and manpower) and the size of each zone audited. Justify chosen routes along transects or through areas. Use GIS to help your decision-making.

Key outcome: describing and justifying the methods and techniques used to collect fieldwork and research data

Fieldwork could involve a range of qualitative and quantitative data linked to assessment approaches. Qualitative data could be field notes, field sketches, photographs/DVDs, extended interviews, focus groups, mobility and accessibility maps. Talk to disadvantaged groups and people trying to overcome a disadvantage. Quantitative data could involve questionnaires (shopping quality, footfall, pedestrian counts, environmental quality assessments, land use, visitor catchment surveys, litter surveys and graffiti assessment).

Use the internet to research geo-demographic data (Acorn and Cameo profiles) plus socioeconomic profiles from National Statistics. Geo-located pictures could help with identity (Flickr, Panoramio, Geograph, etc).

Key outcome: describing and justifying the techniques chosen to present and analyse findings

Your choice of presentation techniques will be influenced by data type. Quantitative data lend themselves to graphs such as line, scatter and histogram, whereas qualitative analysis could use more descriptive narrative techniques (describing a photograph). Some data can be spatially represented (thumbnail pictures of evidence of inequality on a large-scale base map of the study area). Analysis using simple statistics could also be appropriate (mode, mean and median). Inter-quartile ranges could be used for some of the quantitative data collected such as shopping quality and footfalls. Descriptive analysis of qualitative data could be used (open-coding, geographical narratives, abbreviating extended interviews), and a written commentary could accompany a video or photographs of examples of inequality.

Key outcome: commenting on the data and conclusions; evaluating limitations

Examiner tip

Be prepared to write about:

- your fieldwork activities – what you did and what your results showed
- your research – what sources you used and what they showed

Summarise data patterns, trends and anomalies as revealed by analysis of data types (questionnaires, pedestrian flows and densities). Provide a summary of the fieldwork process (particularly the spatial pattern of inequality). How reliable are the findings regarding qualitative and quantitative data? (Too much of one may lead to unreliability.) What other factors could have affected your results? (For example, poor weather when taking photos could make a place look less attractive.) Hidden inequality could exist in rural areas. How does first-hand fieldwork data compare to internet demographics and images of inequality? How could the design process be modified in terms of locations, range of observation techniques, methods of presentation and analysis?

Managing rural inequalities

Overcoming the barriers

Examiner tip

Visit the Commission for Rural Communities website and have a look at their latest State of the Countryside Report, which is an annual survey with lots of good facts and figures.

The hidden and scattered nature of rural inequality means it is often overlooked. According to the Commission for Rural Communities 2007, more than 928,000 rural households are on incomes below the official poverty threshold. The various schemes that address rural disadvantage aim to improve three particular areas.

Financial poverty

This is associated not only with people who are unable to work, but also with people in work. Under-employment (working intermittently) and seasonal employment are more common in rural areas than elsewhere, particularly where agriculture or tourism provide most of the jobs. Financial problems in rural areas mostly affect older people, with pensioner households accounting for around a quarter of those in poverty.

Access poverty

Transport plays an essential role in allowing people to access employment, education, health services, shopping and leisure facilities. Despite generally high levels of car ownership (65% of adults), transport-related disadvantage is more widespread in rural areas. In recent years, ICT has enabled people to participate more fully in society, though country areas often suffer from restricted services such as low or no broadband access. Low-income households in rural areas use digital television and interactive services less than those on higher incomes.

Network poverty

Networks include family, friends, social facilities and local community support. However some rural social networks are breaking down because of:

- the lack of affordable housing
- an increase in commuting (less free time and involvement in the community)
- the forecast growth in mental health problems
- the increase in single-person households
- the growth of ICT methods of working, rather than face-to-face contact

Case Study

Affordable homes for the future

Affordable housing has a crucial role to play in creating better, more sustainable rural communities. In its Housing Green Paper 2007, the government pledged to build 240,000 new houses a year, including 70,000 affordable homes. The vision is to provide access to a decent home for everyone at an affordable price and in a place where people want to live and work.

Strong rural economies underpin the success of schemes to reduce rural disadvantage. The 2007 foot-and-mouth epidemic plus the summer floods, for example, showed how rural communities and economies were vulnerable to environmentally triggered disruptions. Farm **diversification** can be an important method of managing rural inequalities (see Figure 14 on p. 48).

Tourism is the world's fastest growing industry. **Ecotourism** is a way of reducing rural inequalities, particularly in poorer parts of the world. It follows certain principles, for example minimising environmental impact, building environmental and cultural awareness and respect, and providing positive experiences for visitors and hosts. Ecotourism is especially prominent in Kenya, where reducing inequality in areas such as the Masai Mara wildlife reserve depends on:

- partnerships between landowners and tourism operators
- ensuring that revenue generated from tourism reaches community groups and individuals
- using tourism income to improve local infrastructure, such as roads, health facilities and schools

Knowledge check 30

Give a definition of the term 'post-production' agriculture.

Examiner tip

Be prepared to write about the success of different rural schemes to manage problems of inequality. It might be a good idea to practise a 10 mark answer comparing two very different localities and approaches.

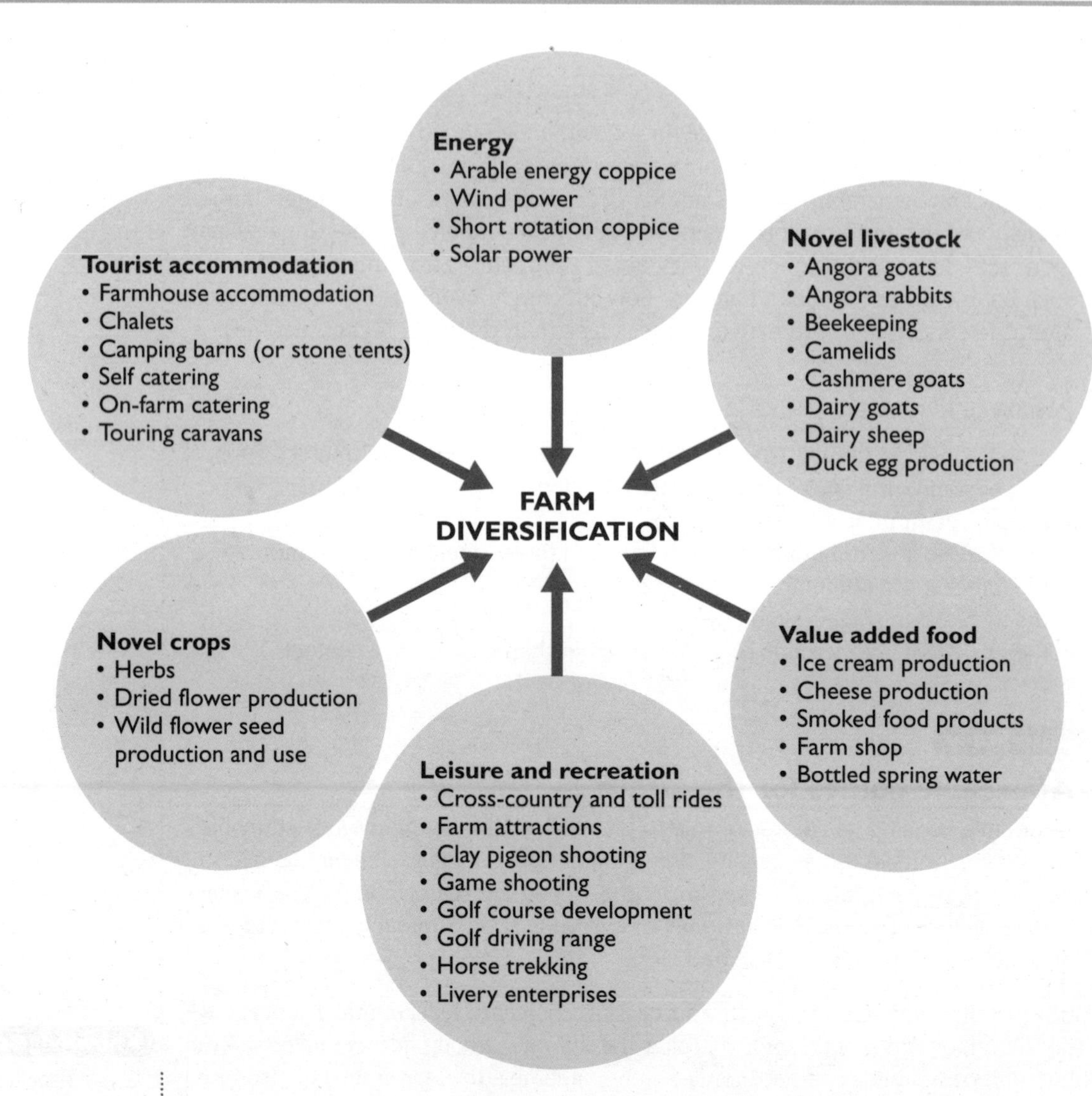

Figure 14 How farms can diversify

Fieldwork and research into examples of rural schemes to reduce inequality

A number of research statistics can help evaluate the success of particular schemes. These include:

- the percentage of households with no car
- the level of unemployment
- the child poverty rate
- the overall poverty rate
- the percentage of people aged 18 and over with no post-school qualifications
- the percentage of households with no central heating
- the illness/disability rate

These census data should be available from the Office for National Statistics. If possible, you should compare pre-scheme and post-scheme statistics.

Knowledge check 31

What is meant by an 'index of multiple deprivation'?

Your fieldwork could include:

- a functional survey
- photographic evidence of change
- interviews with focus/community groups
- questionnaires
- examining local blogs and forums

Another idea is to use **perception analysis**. This can be achieved using prepared statements (see Table 13) directed towards visitors or users. These ideas can be used to evaluate social, economic and environmental impacts.

Table 13 Example of a blank perception analysis chart

Examples of a range of customisable statements	Agree		Disagree	
	Strong +2	Slight +1	Slight –1	Strong –2
Scheme will encourage a range of new employment opportunities for local people (E)				
Scheme will bring much-needed positive change (S, E, En)				
Scheme will benefit local people overall (S, E, En)				
Scheme will encourage the development of new facilities and services for the area (S, En)				
Scheme will bring a better environment for local people (En)				
Local people will find themselves edged out by incomers as a result of the scheme (S)				
New housing associated with the proposal is likely to benefit local people (S)				
The project represents value for money (E)				
The environmental effects will have a negative impact on local ecosystems and biodiversity (En)				
Local people support the proposals (S, E, En)				

Note: E = economic, S = social, En = environmental; there could also be some 'don't knows'

Managing rural inequalities — exam review

Key outcome: designing a fieldwork and research activity to examine the success of rural strategies

You need to select contrasting sites and areas for a complete audit. Justify your chosen location (remote versus accessible rural, or villages that might have different solutions). A series of villages that are handled by different administrators could be surveyed using a common investigation framework. Use GIS to help in decision making, and consider issues of accessibility and proximity. You should refer to different types of sampling such as systematic, stratified or random (for interviewing and questionnaires).

Key outcome: describing and justifying the methods and techniques used to collect fieldwork and research data

Fieldwork could involve a range of qualitative and quantitative data linked to assessment approaches. Qualitative data could be field notes, field sketches, photographs/DVDs, extended interviews, and focus groups to examine change. Talk to community leaders. Quantitative data could be a survey (shopping quality in larger villages, environmental quality assessments, drop-and-collect questionnaires and a dedicated website for responses).

Use the internet to research geo-demographic data (Acorn and Cameo profiles), and socioeconomic profiles from National Statistics. Geo-located pictures could help with identity (Flickr, Panoramio, Geograph etc).

Key outcome: describing and justifying the techniques chosen to present and analyse findings

Your choice of presentation techniques will be influenced by data type. Quantitative data lend themselves to graphs such as line, scatter and histogram. Qualitative analysis could use more descriptive narrative techniques (describing a photograph, for example). Some data can be spatially represented (mini-pictures of evidence of changes in a village on a large-scale base map of the study area). Analysis using simple statistics could also be appropriate (mode, mean and median). Inter-quartile ranges could be used for some of the quantitative data collected, such as shopping quality, diversity, footfalls etc. Descriptive analysis of qualitative data could be used (open-coding, geographical narratives, abbreviating extended interviews, conceptual frameworks), and a written commentary could accompany a video or photographs of change through a timeline.

Key outcome: commenting on the data and conclusions; evaluating limitations

Summarise data patterns, trends and anomalies as revealed by analysis of data (functional change, photos, interviews etc.). Provide a summary of the fieldwork process. (What aspects of rural inequality have been reduced? How effective and manageable are the criteria to measure inequality? How reliable are the findings balancing qualitative and quantitative data? Too much of one may lead to unreliability.) What other factors could have affected results? (Poor weather could mean fewer visitors and affect data.) How does first-hand fieldwork data compare to internet demographics and images of a rural area? Consider how the design process could be modified in terms of sample locations, sample size, methods of presentation, and analytical tools and techniques.

Managing urban inequalities

Causes of and solutions to urban inequalities

Social, economic and environmental inequalities occur in all urban areas. Enormous contrasts in wealth, opportunity, deprivation and exclusion can be found, often virtually side-by-side. However, in urban areas the wealthy and the poor are usually socially segregated. The three main reasons for this are:

- Housing — wealthier people have flexibility of choice of location (near good schools and services), while poorer groups may have a limited choice and might be forced to live in areas of less desirable reputation.
- Ethnic groups, who can suffer discrimination in the job market and are often in low-paid temporary work or unemployed. Newly arrived migrants may not be able to afford anything other than very cheap housing (inner city terraces), which can lead to a concentration or clustering of ethnic groupings. This may also be due to cultural factors. Figure 15 summarises the factors leading to segregation.
- Lower paid groups could find proximity to employment an important factor (especially when served by public transport).

Knowledge check 32

What is meant by the term 'ethnic segregation'?

Internal factors — those within ethnic groups that encourage segregation

New arrivals need mutual support from friends, relatives and community organisations

These immigrants will be happiest with religious centres, ethnic shops and foods, and banks grouping together to serve them

These immigrants need support from areas speaking their own language — a minority language in their new country

These ethnic groups encourage friendship and marriages and reduce contacts, except via schools, which might undermine the culture and traditions of the ethnic groups

Employment and accomodation can often be obtained via networking in an ethnic community

A closely knit ethnic community provides security against abuse and racist attacks — safety in numbers

Ethnic groupings help political power and influence development

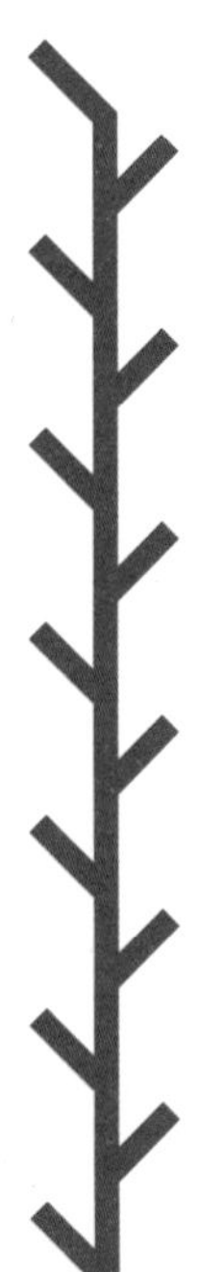

External factors — those within the country or area that encourage ethnic segregation

As immigrants move in, the remaining majority population moves out — in fear of factors such as falling house prices

The majority population is generally hostile or unfriendly to new arrivals

Racism, abuse, racially motivated violence against ethnic minorities or fear of such actions

Discrimination in the job market — ethnic minorities in low-paid jobs or unemployed are forced by circumstances into cheap housing areas and substandard services

Discrimination by house sellers, estate agents and housing agencies keeps ethnic minorities in their ghettos

Discrimination by financial institutions forces ethnic minorities to use their own networks for small business development etc.

Ethnic segregation in urban areas leading to concentration

Figure 15 Factors leading to segregation

Understanding the management of urban inequalities requires an overview of the issues in urban environments. These are summarised in Table 14.

Table 14 Issues affecting urban environments

Economic	Lack of jobs, linked to loss of industry, provision of traffic systems, public transport, retail opportunities and services
Social	Deprivation and poverty, segregation, ageing society, housing and health, education and training opportunities
Environmental	Sustainability, pollution control, ecological and carbon footprints, heritage and conservation plans

There are also issues associated with the image of a town or city, i.e. its reputation, the amount of investment and how it is governed.

Various groups, organisations and individuals can be involved in reducing urban inequalities. These are often referred to as 'players' or 'stakeholders'. The spectrum of stakeholders is illustrated in Figure 16.

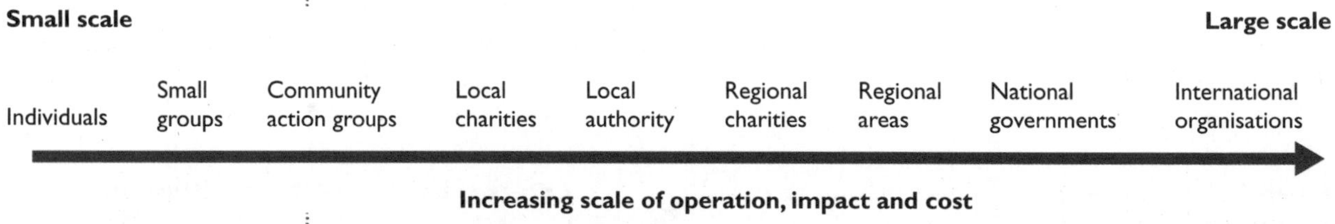

Small scale **Large scale**

Individuals | Small groups | Community action groups | Local charities | Local authority | Regional charities | Regional areas | National governments | International organisations

Increasing scale of operation, impact and cost

Figure 16 Stakeholders involved in reducing urban inequalities

Examiner tip
For your case study on urban inequalities, know the names, roles and influence of various stakeholders in relation to particular schemes.

Managing urban inequality has a number of strands:
- changing planning laws to encourage development of new city/urban sites and reducing inequalities in health ('postcode lotteries')
- social housing schemes, affordable public transport, tackling low-pay issues (minimum wage)
- joined-up policies at local level, with empowerment of local residents and communities

Case Study

Reviewing the sustainable communities plan

The £38 billion sustainable communities plan (2003) set out an ambitious and long-term vision for creating thriving and sustainable communities in all regions of the UK. The main emphasis was on delivering more decent and affordable homes and ensuring the right infrastructure — that schools, hospitals, shops and green spaces were in place to create areas where people would want to live and work, now and in the future.

In the urban context there are a number of successful examples:
- The increase in sustainable development across southeast England, particularly the Thames Gateway, by incorporating new housing with high quality design and environmental sustainability, plus good transport links, services, leisure facilities and open spaces.
- In the north and Midlands, £1.2 billion has been invested over 5 years in nine pathfinder areas. In many cases this means a programme of housing-market renewal to bring run-down housing stock up to modern-day standards.

- In the first 10 years of the scheme the government supported the creation of over 230,000 new affordable homes in urban areas. Most are for rent or for sale through low-cost home ownership schemes.

Fieldwork and research into examples of urban schemes to reduce inequality

It is difficult to evaluate the success of individual schemes. Fieldwork and research should be aimed at building up 'before' and 'after' pictures of an urban area. It then becomes possible to make judgements on particular schemes.

There are a range of local primary fieldwork opportunities:

- Mapping the distribution and location of security cameras in an urban area. Camera hotspots can be linked to facilities that might be open late at night (pubs, bars, fast food outlets).
- The distribution of Neighbourhood Watch stickers and evidence of neighbourhood policing.
- Mapping the distribution of gated communities in an urban area. You could examine the evolution of such neighbourhoods — are they adjacent to rich or poor areas? What are the reasons for this?
- Street cleanliness could reveal something about the quality of the environment and could be used as an indicator of a scheme's success.
- Accessibility is another component of inequality that can be researched. The 1995 Disability Discrimination Act gave disabled people rights in employment, education, access to goods and services, and buying or renting land or property. You could complete a survey of car parking in the local area and compare the views of disabled users with those of able-bodied users.
- Interviews with key players (town centre managers, planners, local charity workers or recipients of support) can also provide information on particular schemes.
- Photographs can be extremely useful in telling the story of an area.

Examiner tip

Be able to describe in some detail how you could use photographs to 'tell the story' of an area.

Managing urban inequalities — exam review

Key outcome: designing a fieldwork and research activity to examine the success of urban strategies

Select sites and areas for a complete audit, and justify your chosen locations (contrasting schemes in one large urban area or different schemes in different towns of contrasting size, type, location, identity). Use GIS to help in your decision-making. You should refer to different types of sampling, such as systematic, stratified or random for questionnaires, and consider the availability of manpower to work different areas to complete tasks.

Key outcome: describing and justifying the methods and techniques used to collect fieldwork and research data

Fieldwork could involve a range of qualitative and quantitative data linked to assessment approaches. Qualitative data could be field notes, field sketches, photographs/DVDs, extended interviews, and focus groups to examine change. Talk to community leaders, town centre planners, community police and schools. Quantitative data could be customised environmental quality assessments on the street, drop-and-collect questionnaires or a dedicated website for responses, plus mapping of new facilities and schemes including CCTV maps. Use the internet to research geo-demographic data (Acorn and Cameo profiles) and population change (from National Statistics). Also use geo-located pictures to help with identity and change (Flickr, Panoramio, Geograph etc.). Evidence could come from blogs/forums and community sites that examine success.

Key outcome: describing and justifying the techniques chosen to present and analyse findings

Choice of presentation techniques will be influenced by data type. Quantitative data lend themselves to graphs such as line, scatter and histogram, whereas qualitative analysis may use more descriptive narrative techniques (describing a photograph illustrating change). Data can be spatially represented (mini-pictures of changes in towns on a large-scale base map of the study area). Analysis using simple statistics could be appropriate (mode, mean and median). Inter-quartile ranges could be used for some quantitative data collected, such as environmental quality. Qualitative ways of analysing data could be more descriptive (open-coding, geographical narratives, abbreviating extended interviews, conceptual frameworks, and a written commentary which could accompany a video/DVD or a series of pictures illustrating change through a timeline).

Key outcome: commenting on the data and conclusions; evaluating limitations

Summarise data patterns, trends and anomalies as revealed by analysing the data (functional change, photos, interviews etc.). Complete an overall summary of the fieldwork process, showing what aspects of urban inequality have been reduced and for whom. Has one scheme been more successful than another? How reliable are the findings in terms of balancing qualitative and quantitative data? (Too much of one may lead to unreliability.) What other factors could have affected results? (Poor weather could mean fewer visitors.) How does first-hand fieldwork data compare to internet demographics and images of the urban area? Consider how the design process could be modified in terms of locations, sample size, techniques, methods of presentation and analytical tools.

Summary

Unequal spaces

- Recognise that inequality has a spatial dimension and operates at a wide range of scales.
- Appeciate that inequality may manifest itself through social, technological, and environmental inequalities.
- Know how to design a survey to look at the patterns of spatial inequality in both urban and rural areas.
- Know how research sources can be used to explore both patterns of inequality and look for evidence of managing success.
- Appreciate that the causes of inequality in rural and urban areas will be different.
- Appreciate how creating a tourism brand may be an important part of rural and urban solutions to inequality.
- Recognise that solutions to urban and rural inequalities may overlap heavily with rebranding strategies.

Rebranding places

Time to rebrand

What is rebranding and why is it needed in some places?

Regeneration is a buzzword that has been applied to a wide range of contexts. In this context it refers to the physical redevelopment of landscapes — in city centres or rural areas — with the intention of promoting economic development by attracting external investment. This investment includes jobs (company relocations), money (leisure and business tourists) and people (wealthy residents attracted to lavish city-centre apartments or countryside barn conversions).

Re-imaging is how cities construct and promote positive images of themselves. Promoting the right image can be crucial to the success of a regeneration scheme.

Rebranding, the focus of this section, is improving a place's image and people's perception of it. In other words, helping to sell a place to a target audience. The most dramatic change in many cities since the 1980s has been the redevelopment of the city centre. Many parts of rural landscapes (from remote to more accessible) have also undergone transformations, with the diversification of agriculture and a move towards leisure, pleasure and cultural industries.

Knowledge check 33

Examine the difference between cultural and heritage rebranding of places.

Attracting people through rebranding

Rebranding (Figure 17) and re-imaging are processes that make areas that were in decline attractive to a range of potential customers. The actual process is often costly and disruptive, as it involves the physical redevelopment of worn-out or outdated facilities and infrastructure. Rebranding and re-imaging are also concerned with a place's reputation, spirit and identity.

How can rebranding be achieved?

There are various strategies that help places reinvent themselves to become attractive to visitors, residents and people seeking to invest. Often there is a focus on four distinct areas:

1. **social:** to overcome inequalities, deprivation and poverty
2. **economic:** to improve job opportunities and attract inward investment
3. **environmental:** to improve the general environment, such as the removal of derelict buildings
4. **political:** using the bid industry (lottery funding) to generate income

Marketing can attract visitors and start the regeneration process. This will involve promotional materials such as brochures and leaflets, videos, CDs, websites, newspaper and magazine advertisements, slogans and logos. Marketing can help a place promote its unique selling point. Figure 18 shows how linked factors attract visitors to places.

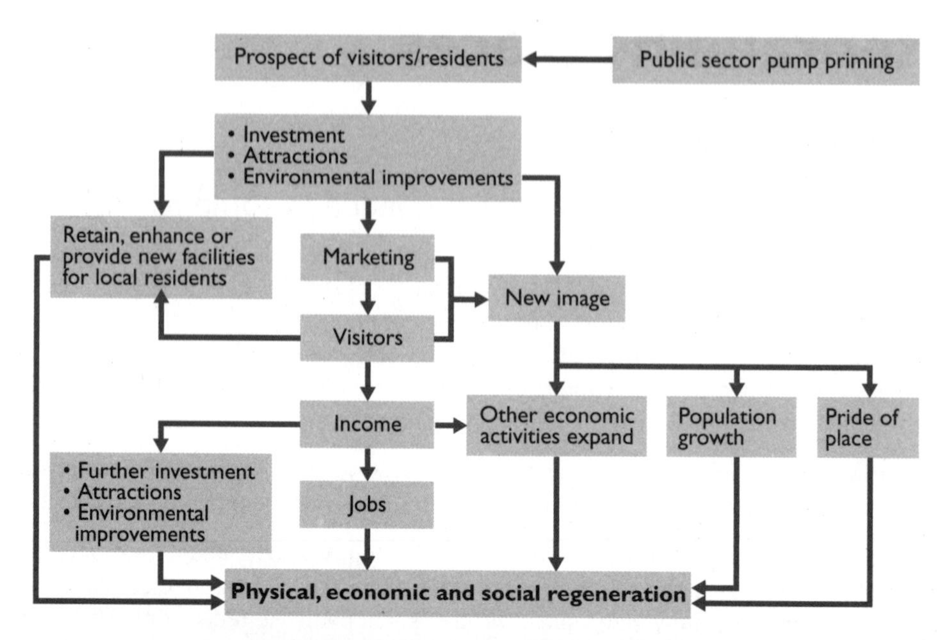

Figure 17 The strategy for rebranding

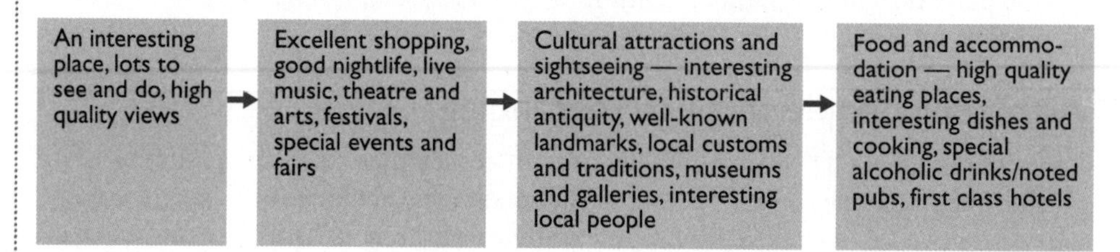

Figure 18 Marketing a location

Knowledge check 34

With reference to Figure 17, what is meant by the term 'public sector pump priming'?

Case Study

Rebranding Notting Hill

Creative industries generate around £21 billion per year and employ approximately 500,000 people in London. They include businesses covering architecture, design, publishing, music, fashion and new manufacture (such as hand-made jewellery), as well as television and film.

The film *Notting Hill* is an example of how a movie associated with a particular area can be a catalyst for improvement. *Notting Hill* portrayed the area as fresh and exciting. The film is also credited with helping to make the area one of London's most fashionable districts.

Why is rebranding necessary?

Different places have varying needs in terms of rebranding. Reasons why somewhere might need to be rebranded could include:

- **the economy:** loss of dynamism in the economy and a lowering of the tax base — often linked to the loss of mainstay industries such as coal mining or steel production

- **the environment:** where 1960s-style planning has resulted in areas built for the car and concrete buildings that are now described as dirty or ugly
- **image:** vital for economic development, especially tourism and inward investment

Table 15 How and why areas have declined

Towns and cities	Countryside	Coastal areas
• Loss of retailing provision (especially food, electrical and DIY) to out-of-town locations • Loss of business and commercial functions to more peripheral locations such as science parks, and loss of industries in some towns and cities • Inaccessibility for the private motorist (congestion charging, parking) and the expense of getting to the city centre • Increasing costs — might have to employ a town centre manager to promote and market the area	• Lack of transport infrastructure; public transport infrequent and car ownership expensive • Agricultural change — mechanisation, competition from overseas, high-profile diseases (e.g. foot and mouth) • Moving away from government-supported agriculture; increased requirement to diversify to survive • Depopulation in some areas — loss of residents due to limited employment opportunities and expensive housing (especially for first-time buyers)	• Ease and affordability of overseas travel makes UK coastal areas appear less attractive and plagued by bad weather • Decline in the hotel and guesthouse trade — rundown appearance and decline in resort economy • Difficulty of attracting private investment • Continued inaccessibility of some coastal towns while transport infrastructure has been improved in many other parts of the country • Possible decline in traditional fishing industries

In addition to the above, former industrial towns could suffer from the decline of traditional manufacturing industries and a loss of jobs. These areas may be characterised by dereliction and contamination (mostly the legacy of mining). There would not be a strong tradition of self-employment or business start-ups, there could be substantial education and training deficits, and the areas may have high concentrations of migrant labour engaged in low-paid employment.

Changes in tourism and leisure have brought a series of winners and losers and have highlighted the need for rebranding at a range of locations.

Table 16 Winners and losers in leisure and tourism

Winners	Losers
• Overseas destinations • Near-motorway locations • Branded hotels • Self-contained holiday villages • High-quality independent hotels • Night-time economy • Cities	• Visitor attractions • Traditional seaside resorts • B&Bs and guesthouses • Mid/lower-market hotels and guesthouses • Peripheral locations

The main objectives of re-imaging and rebranding are improvements to:

- the physical environment, by focusing on sustainability
- the quality of life of populations, by providing better living conditions and cultural activities
- social welfare
- the economic prospects of populations, through job creation, education or re-skilling programmes

Uncovering the profile of places in need of rebranding

This is a core fieldwork and research item. The specification refers to the profile of a place — a description of what the place is like — similar to an identity audit.

A range of surveys and secondary research sources can be used to establish a profile. A chart resembling Table 17 can be used to start the auditing process.

Examiner tip
Questions relating to the profile of places will require you to do both fieldwork and research. The research could be focused on ethnographic and geo-demographic data, together with looking into how the place has changed, e.g. historical photos and postcards.

Table 17 Starting the auditing process — an example for one location

Site/location	What we like	What we dislike
Next to the post office	Views of countryside	Bad language from some people
	Wide pavements	Fast cars passing close-by
	Busy with people coming in and out	Smell of dustbins
	Community feel	
	Good upkeep of street and furniture	

A variety of environmental surveys could be employed (street quality, shopping quality, landscape quality, litter etc.).

Time to rebrand — exam review

Key outcome: designing a fieldwork and research activity to find out why rebranding is necessary

Select sites for a complete audit by uncovering the profiles of areas. Justify chosen locations (rural, urban, coastal) and look for a history of problems (change, neglect). Check for proximity to school or field centre. You should refer to different types of sampling such as systematic, stratified or random.

Key outcome: describing and justifying the methods and techniques used to collect fieldwork and research data

Fieldwork could involve a range of qualitative and quantitative data linked to assessment approaches. Qualitative data could include field notes, field sketches, photographs, extended interviews, focus groups and customised place-check forms. Quantitative data could be a range of questionnaires (retail, shopping quality, footfall, pedestrian count, and other personalised environmental quality assessments such as litter and graffiti).

Use the internet to research geo-demographic data (Acorn and Cameo profiles), socioeconomic profiles from National Statistics, and geo-located pictures to help with identity (Flickr, Panoramio, Geograph etc.).

Key outcome: describing and justifying the techniques chosen to present and analyse findings

Your choice of presentation techniques will be influenced by data type. Quantitative data lend themselves more to graphs such as line, scatter and histogram.

Qualitative analysis could use more descriptive narrative techniques (describing a particular photograph). Some data can be spatially represented (thumbnail pictures of areas needing to rebrand on a large-scale base map of the study area). Analysis using simple statistics could also be appropriate (mode, mean and median), and inter-quartile ranges could be used for some of the quantitative data collected (retail quality, diversity, footfalls etc.). More descriptive qualitative analysis could include open-coding, geographical narratives, abbreviating extended interviews, conceptual frameworks, and a written commentary to accompany a video/DVD or series of pictures depicting urban decay to justify the need for rebranding.

Key outcome: commenting on the data and conclusions; evaluating limitations

Summarise data patterns, trends and anomalies as revealed by analysis of the range of data (questionnaires, pedestrian flows and densities). Provide a summary of the fieldwork process (does the area studied need a rebrand?). How reliable are the findings in terms of the balance of qualitative and quantitative? (Too much of one could lead to unreliability.) What other factors could have affected results? (Poor weather when taking photos could make a place look less attractive.) How do first-hand fieldwork data compare to internet demographics and images of place identity? How could the design process be modified in terms of locations, sample size, methods of presentation and analytical tools and techniques?

Rebranding strategies

Who are the rebranding players and what strategies exist for places to improve themselves?

Successful regeneration requires an ability for risk-taking and clear entrepreneurial management. Baltimore (USA) was used as a model regeneration example in the 1980s, when solutions coming from North American policy ideas were based on free-market principles. The secret was combining functions, processes and catalysts, and diversifying the economy:

- residential
- commercial
- culture
- sport
- tourism (business and individual)

Since the 1980s policy-makers in the UK have increasingly used the partnership model of rebranding and regeneration, by including community and voluntary sectors alongside local authorities and the private sector. Often these partnerships have been established as '**bottom-up**' approaches including progressive planning policies, community economic development and community architectural schemes. This is in contrast to '**top-down**' enterprises that come from central government and tend to establish hierarchies of power among stakeholders. Bottom-up approaches are based on listening to local opinions and devising local solutions.

Examiner tip

For your case study on rebranding, know the names, roles and influence of various stakeholders in relation to particular schemes.

Knowledge check 35

What is meant by the term 'knowledge industries' in Figure 19?

Stakeholders are often pivotal in the success or failure of rural and urban schemes. Those most influential to regeneration and re-imaging programmes (and most affected by them) are central government, local government, the private sector, community and voluntary sectors, and local residents. Rebranding strategies may use three types of intervention, as Figure 19 shows.

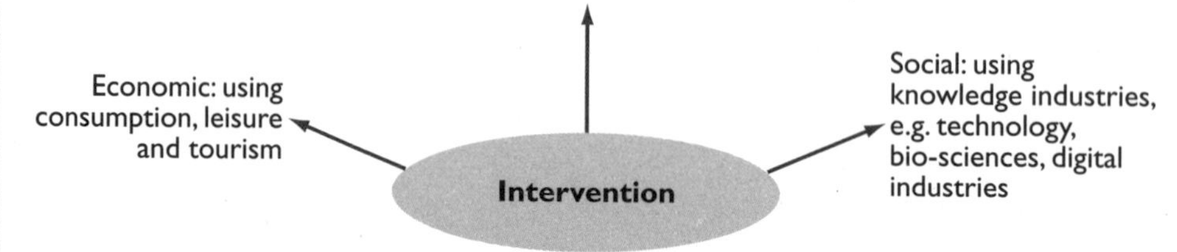

Figure 19 Three types of intervention

Case Study

Key players in the 2012 Olympic Games

The 2012 Olympic Games in London will be a massive sporting occasion which will regenerate the city's run-down East End. In terms of stakeholders, the picture is complex, with a number of organisations involved — see Table 18. The overall success of the project, in terms of the local economy, people and environment, will be affected by the decision makers of these organisations.

Table 18 How the 2012 Olympic Games will be run

Stakeholder	Role
International Olympic Committee	The IOC cooperates with the UK organisations responsible for the delivery of the games
UK government	Sets up a range of organisations, including the London Development Agency, London Organising Committee of the Olympic Games, and the Olympic Delivery Authority. These organisations are responsible for economic regeneration, planning the games and building facilities
London Assembly	Comprises Boris Johnson (Mayor of London) and Transport for London, and will organise the movement of 500,000 people per day during the games. The assembly will also influence policy
Local government	Four London borough councils are affected by the Olympics. They have considered and approved the various Olympic planning applications for housing, transport and infrastructure

Case Study

Making a seaside resort attractive to visitors

Coastal towns can be made attractive to a range of user groups in a number of ways:

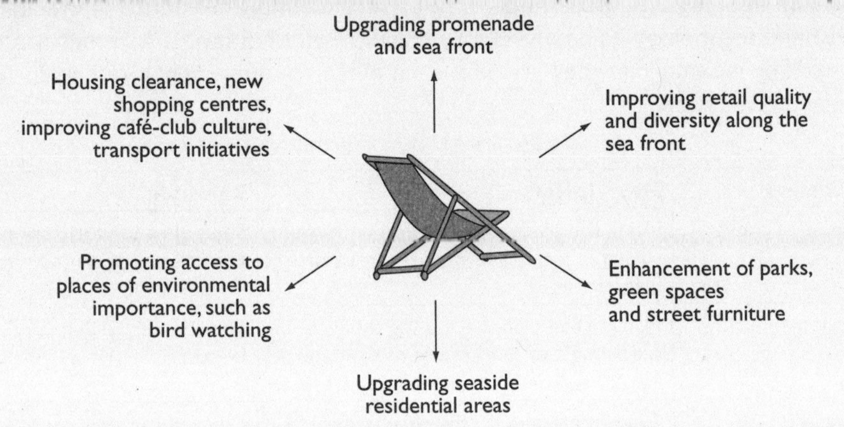

Examples of resorts that have taken this approach include Tenby, Great Yarmouth and Brighton.

Fieldwork and research into rural strategies

The countryside can rebrand itself through a range of different activities, often by playing to its strengths. This could be marketing products that are special or unique to an area and make it different from a neighbouring resort. Table 19 gives examples of rural rebranding and where this has been successfully applied.

Table 19 Examples of rural rebranding

Strategy	Description	Example
Community radio	Local non-commercial radio station serving a small area. Could be linked to a local event such as an arts festival	10 Radio, Somerset
Food festival	A celebration of local food. Attracts many visitors to an area during the festival window	Ludlow Food Festival
Rural sports	Quad biking, shooting, fishing, paint-balling and other activities to attract visitors to an area. Could be a farm diversification scheme	Torbay Fishing Festival
Food trail	A free map detailing good places to eat. Visitors are attracted at all times of the year	The Ribble Valley Food Trail
Media and film-making	Film and television trails are increasingly popular ways of getting people to explore more remote rural landscapes that have been used for film and television production	Movie Map, North Wales
Farm diversification	New agricultural enterprises to increase farm incomes — 40% of UK farm income comes from diversification	Runnage Farm, Dartmoor provides bunk/barn accommodation
Rural industries	A range of knowledge-based industries and specialist services, e.g. furniture-making	Jim Lawrence Traditional Ironwork, Suffolk

Fieldwork and research into urban strategies

Knowledge check 36

What is 'signature architecture' in the context of urban rebranding?

Urban areas have a range of rebranding strategies or themes available (Table 20). In line with rural regions, urban areas draw on their identity and culture. Many cities use sport and technology as catalysts for regeneration. High profile schemes are made to work by making sure they are part of an overall improvement strategy, including social regeneration.

Table 20 Rebranding strategies

Strategy	Description	Example
Cities of sport	Using the place's connection with sport to act as a catalyst for further inward investment	Sheffield and world snooker Manchester: Commonwealth Games
Sustainable cities	Many urban places are using their green credentials to market themselves and attract a variety of people and investors	Manchester Green City Leicester
Specialist industries	Cities may attract new small-scale businesses and enterprise, especially knowledge industries	Custard Factory, central Birmingham Cambridge Science Park
Food cities	Many cities are highlighting their range of high quality global cuisine to attract visitors	Taste of Birmingham
City of culture	Using culture as a catalyst to attract new people and associated businesses to an area, e.g. café-club scene	Liverpool as City of Culture 2008
Arts and heritage	A wide range of visitor attractions (historic buildings, museums, galleries etc.) give the town or city a quality feel	Morecambe/Lancaster
Use of innovative architecture	Design of a new building could be a competition. Signature architecture can help with a town or city's identity	Tate Modern, London Hilton, Manchester

Rebranding for a sustainable future

Many rebranding strategies accommodate elements of sustainable development. This can be depicted as a three legged stool (Figure 20).

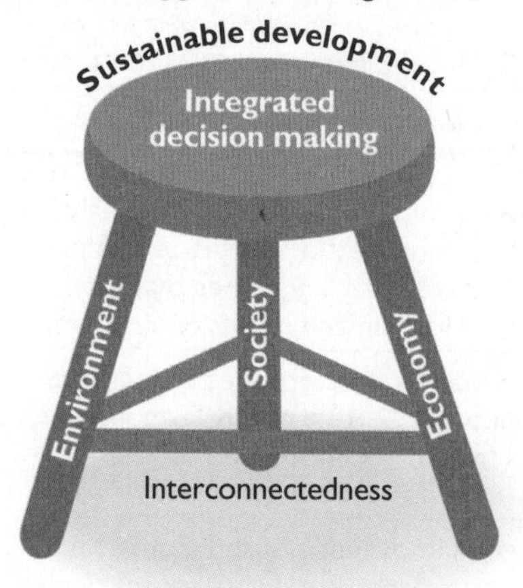

Figure 20 A model of sustainable development

Case Study

Barcelona's sustainable energy solutions

Barcelona's plan for energy improvement (2002–10) aims to increase the use of renewable energy (especially solar energy), reduce the use of non-renewable energy sources, and lower the level of emission gas produced. The plan could produce an environmentally sustainable city, cutting air pollution and reducing the use of fossil fuels. New buildings and those undergoing refurbishment are required to use solar energy to supply 60% of their hot water. The scheme produces annual savings of around 12,000 megawatt hours and a corresponding reduction in carbon dioxide emissions of about 2,000 tonnes per year.

Examiner tip

The Barcelona case study is a good length for integration into a 10 mark question knowledge/case study framework. You may also be asked to comment on its success, so some additional research would be a good idea.

Rebranding strategies — exam review

Key outcome: designing a fieldwork and research activity to look at players and strategies

Fieldwork and research design for this part of the specification could focus as much on secondary evidence as on primary evidence. The design will require setting up contacts (main players or stakeholders) for urban and rural contexts. The contacts should provide valuable information on the reasons for, and the strategies involved in, rebranding. Fieldwork and research will be combined when evaluating schemes.

Key outcome: describing and justifying the methods and techniques used to collect fieldwork and research data

Fieldwork could involve qualitative data (notes, recordings, transcripts of interviews with players or stakeholders). Quantitative data could be questionnaires linked to spheres of influence or changes in perception, shop footfalls, environmental quality (in particular graffiti, street quality, retail quality, retail types and occupancy rates).

Research secondary evidence of success in rebranding (e.g. photos illustrating change, such as old postcards or historical websites, GIS maps of employment, and socioeconomic statistics). Use blogs, YouTube, MySpace and Flickr to uncover identity as reflected by internal and external users. Geo-demographic data using postcodes could be available, and changes in rateable values (see the Valuation Office Agency website) could be indicators of success.

Key outcome: describing and justifying the techniques chosen to present and analyse findings

Choice of presentation techniques will be influenced by data type. Quantitative data lend themselves to graphs and cartography (retail quality, comparisons of shop types). Qualitative analysis may use more descriptive narrative techniques to annotate change in a particular urban scene (old versus new). Some data can be spatially represented (e.g. thumbnail pictures of rebranded flagship architecture on a large-scale base map of the urban zone).

Analysis of descriptive statistics could also be appropriate (mode, mean and median). Inter-quartile ranges could be used for some of the quantitative data collected, such as shop footfalls. Other techniques could be more descriptive (open-coding, geographical narratives, abbreviating extended interviews, conceptual frameworks, and a written commentary to accompany a video/DVD transect through an urban area). Textual analysis of promotional websites and literature could pay dividends.

Key outcome: commenting on the data and conclusions; evaluating limitations

You should summarise data patterns, trends and anomalies as revealed by the analysis of the range of data (questionnaires, activity patterns, environmental quality). You should also present a summary of the fieldwork process. What evidence is there of successful rebranding in rural areas? How reliable are the findings in terms of the balance of qualitative and quantitative data? (Too much of one may lead to unreliability.) What other factors could have affected results (poor weather leading to low visitor numbers, or out of season)? Environmental quality could be subjective unless pre-calibrated. How do first-hand fieldwork data compare to internet images of rural rebranding? How could the design process be modified in terms of sample location, sample size, methods of presentation, and analytical tools and techniques?

Managing rural rebranding

How successful has rebranding been in the countryside? Fieldwork and research into the success of specific rural rebranding examples

Case Study

Rural woman diversifies local economy

Ex-teacher and farmer Caroline Hubbard set up the Ginger Piggery — a food and art centre — as a new venture on the family farm at Boyton, near Warminster in Wiltshire. The farm supplies the shop with home-reared meat, produced to a high standard using a herd health plan drawn up by Bristol University. Ginger Piggery also sells meat and food from over 50 producers within an 80 km radius, promoting the value of traceable, high quality local produce. Caroline also runs craft workshops, exhibits work of more than 20 local artists, and runs farm tours and corporate activities. This is a good example of how local distinctiveness has been used to diversify and rebrand agriculture.

Knowledge check 37

What are and why are 'catalysts' important for rebranding?

Case Study

The use of architecture in rural settings to inspire and regenerate

Culture has often been used as a potential catalyst for regeneration in cities, but does it work in rural areas? Recent research has illustrated that architectural symbolism is a powerful tool in terms of spearheading the regeneration process, both for the countryside and for towns. A good example is Cornwall's Eden Project.

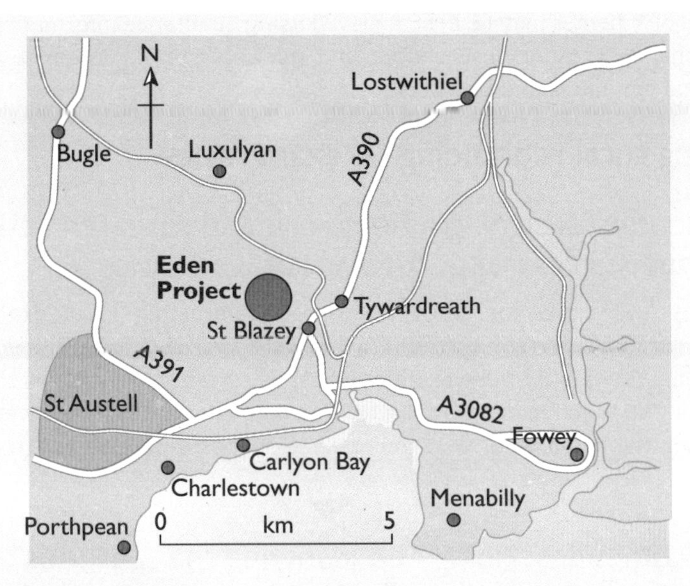

Figure 21 Location of the Eden Project near St Austell, Cornwall

Broadening the definition of culture beyond museums and arts centres brings the economic achievements of the Eden Project — perhaps one of the UK's best-known attractions — into focus. By March 2001, it had attracted 1.91 million visitors, having anticipated only 200,000 in 2000 and 750,000 in 2001. Visitor numbers have now stabilised at about 750,000 a year. The Eden Project is one of the prime attractions for visitors to Cornwall. Tourists spend £5 million a year at the site, with a further £4.3 million being spent off-site.

The Eden Project employs around 600 permanent staff. 95% were recruited locally, and 50% were previously unemployed. In addition, Eden's purchasing plan has secured 197 jobs in supplier businesses, of which 33 are in the local St Austell area. The Eden brand has become the symbol of regeneration in the locality. The project has not been without problems. Traffic and congestion have increased, especially during busy holiday periods. As most visitors arrive by road this has had a knock-on effect of increasing pollution in the area. Attempts to get visitors to use public transport have met with limited success.

Examiner tip

Using the information in the Eden case study, create a simple table that looks at the advantages and disadvantages of this development.

Assessing the success of rural rebranding projects could require a range of economic, environmental and social evaluations. Various techniques can be used to look at these elements, as shown in Table 21.

Table 21 Rebranding evaluations

Evidence	Details and examples
Economic	Visitor rates (compare with historic), footfall counts, population statistics, list of functions and services
Social	Place reputation using extended interviews and open questions, focus groups, interviews with employees/managers. Catchment surveys from questionnaire response
Environmental	Biodiversity, litter surveys, environmental quality, air quality and pollution statistics. Historical maps

Evaluation of schemes may be best achieved using qualitative information (sketches, photographs, participant observation) and the analysis of written texts, such as promotional brochures and websites.

Managing rural rebranding — exam review

Key outcome: designing a fieldwork and research activity to look at the success of particular rural schemes

Select sites and areas for study — choose contrasting rural examples (an accessible rural area versus a remote or discrete area such as a village or larger rural region). You should refer to different types of sampling such as systematic, stratified or random. Discus how and why villages or rural areas were divided up (consider the number of researchers and manpower). Use GIS to help site selection.

Key outcome: describing and justifying the methods and techniques used to collect fieldwork and research data

Fieldwork could involve a range of qualitative and quantitative approaches for each location. Qualitative data could include field notes, field sketches, photographs, extended interviews, focus groups, oral histories, photos and video/DVD, and activity maps (ethnographic fieldwork). Quantitative questionnaires could be linked to sphere of influence, visitor footfalls (in busy locations) and customised environmental quality surveys. Research secondary evidence of the success of rural rebranding (photos illustrating change such as old postcards and historical websites, changes in employment, visitor profiles and published catchment surveys). Use blogs, YouTube, MySpace and Flickr to uncover identity as reflected by both internal and external users.

Key outcome: describing and justifying the techniques chosen to present and analyse findings

Your choice of presentation techniques will be influenced by data type. Quantitative data lend themselves to graphs and cartography (ray diagram for sphere of influence and dot distribution activity maps). Qualitative analysis may use more descriptive narrative techniques (annotating a photograph). Some data can be spatially represented (thumbnail pictures of rebranded facilities on a large-scale base map of the village or region). Analysis of descriptive statistics could also be appropriate (mode, mean and median). Inter-quartile ranges could be used for some quantitative data collected, such as visitor footfalls etc. Other techniques could be more descriptive, e.g. open-coding, geographical narratives, abbreviating extended interviews with focus groups, conceptual frameworks, and a written commentary to accompany a video, DVD, or a series of images (such as pictures of farm diversification illustrating a range of specialist food products).

Key outcome: commenting on the data and conclusions; evaluating limitations

Summarise data patterns, trends and anomalies as revealed through the analysis of the range of data (questionnaires, activity patterns, environmental quality). Provide a summary of the fieldwork process. What evidence in the rural area(s) is there of successful rebranding? How reliable are the findings in terms of balancing qualitative and quantitative data? (Too much of one may lead to unreliability.) What other factors could have affected your results? Poor weather could mean low visitor numbers. Environmental quality can be subjective unless pre-calibrated. How does first-hand fieldwork data compare to internet images of the rural rebrand? How could the design process be modified in terms of sample location, sample size, techniques, methods of presentation and analytical tools and techniques?

Managing urban rebranding

How successful has rebranding been in urban areas? Fieldwork and research into the success of specific urban rebranding examples

The ingredients for a successful town, city or urban area are:

- technology and innovation
- quality education
- sound transport and communication networks
- being culturally and ethnically diverse
- being safe and secure
- having a low-skilled and high-skilled labour mix
- possessing a distinctive character (including retail)
- having historic interest, high quality design, green spaces, social interaction (cafés and bookshops)
- having a mixture of residential tenures and types

Case Study

Urban Splash: agents of rebranding

Urban Splash is the company behind some of the UK's most exciting urban projects. It claims to re-invigorate worn out areas and is involved in the development of new housing (especially the conversion of existing buildings, new offices, new retail units, bars, cafés and hotels). The company's first scheme was the redevelopment of six Victorian mill buildings in Castlefield, Manchester. It was the catalyst for regeneration schemes by other developers totalling £300 million in the city.

To date the company has invested in over £500 million of buildings, more than 85,000 m^2 of commercial floor space, and in excess of £100 million of new homes. Urban Splash is interesting as a case study because it will invest in projects that other developers feel are undesirable. If successful, their projects can be catalysts for further development and regeneration.

Knowledge check 38

In Figure 22 what is meant by (a) oral history and (b) retail diversity?

The focus for fieldwork and research is evaluating the success of rebranding. Figure 22 gives some ideas of success.

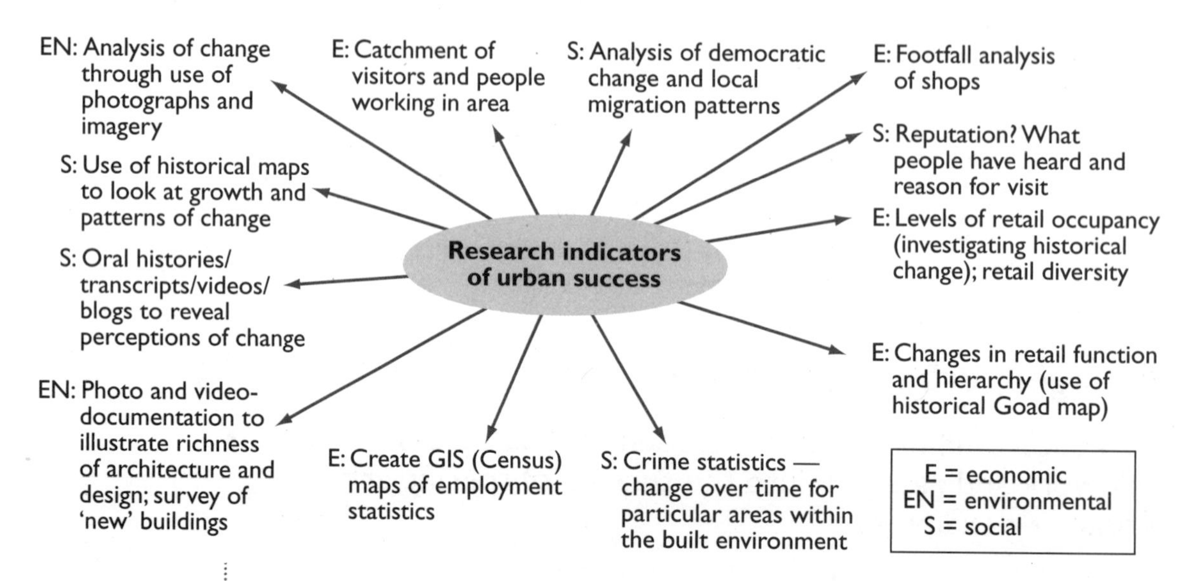

Figure 22 Judging rebranding

Examiner tip

For an exam you should be prepared to comment on the strengths and weaknesses of recording sheets such as Table 22, together with your own recording sheets.

There are a number of possible fieldwork activities. Environmental quality is important, as are footfalls and visitor numbers. Questionnaires could reveal perceptions of success and photos/video/DVD can be used to record visual imagery. Table 22 is a recording sheet that could be used to evaluate environmental and other impacts of rebranding.

Table 22 Evaluating impacts of rebranding

	Negative impact			No impact	Positive impact		
Factors	Strong –3	Medium –2	Slight –1	0	Slight +1	Medium +2	Strong +3
Overall scenic quality of area							
Building design							
Effect on any wildlife habitats							
Effect on land used for economic purposes							
Effect on land used for social purposes e.g. housing							
Visual impact of a building — height							
Noise impact of new development							

	Negative impact			No impact	Positive impact		
Factors	Strong –3	Medium –2	Slight –1	0	Slight +1	Medium +2	Strong +3
Likely impact of new development on local economy							
Value/utility for a range of users							
Amenity value of the area							

Other ideas include:

- investigating the 24-hour city (particularly issues of exclusion and nightlife associated with redevelopment of central areas)
- looking at the degree of cloning in town centres
- investigating sport as an agent of rebranding, including the range and type of any new facilities, reputation of city, perception of change through questionnaires etc.

Managing urban rebranding — exam review

Key outcome: designing a fieldwork and research activity to look at the success of particular urban schemes

Select sites and areas for study to complete the work. You should choose contrasting urban examples or a transect through a larger urban area. Discuss types of sampling (systematic, stratified or random) in conjunction with other methods (questionnaires, environmental quality survey points). Think about how and why urban zones were divided up (number of researchers and manpower). Use GIS to help in site selection.

Key outcome: describing and justifying the methods/techniques used to collect fieldwork and research data

Fieldwork could include qualitative field notes, photographs (especially flagship architecture), focus groups, oral histories, DVD transects through town or city, activity maps (ethnographic fieldwork), and extended interviews with residents and tourists. Quantitative data could be questionnaires linked to a sphere of influence, changes in perception, shop footfalls, and environmental issues (graffiti, street quality, retail quality, retail type and occupancy rates). Research secondary evidence of success (photos illustrating change, such as old postcards and historical websites, GIS maps of employment, and socioeconomic data). Use blogs, YouTube, MySpace and Flickr to uncover identity as reflected by internal and external users. Obtain geo-demographic data via postcodes. You can explore changes in rateable value in recent years by looking at the Valuation Office Agency website. Change may be an indicator of success.

Key outcome: describing and justifying the techniques chosen to present and analyse findings

Choice of presentation techniques will be influenced by data type. Quantitative data lend themselves to graphs and cartography (bar charts of environmental quality assessment and retail quality, comparisons of shop types using historical Goad maps). Qualitative analysis could use more descriptive narrative techniques (annotating change in a particular urban scene — old versus new). Some data can be spatially represented (thumbnail pictures of rebranded flagship architecture on a large-scale base map of the urban zone).

Descriptive statistics could be appropriate (mode, mean and median). Inter-quartile ranges could be used for some of the quantitative data, such as shop footfalls. Other techniques could be more descriptive (open-coding, geographical narratives, abbreviating extended interviews, conceptual frameworks, and a written commentary to accompany a video or DVD). A textual analysis of promotional websites and literature could be carried out.

Key outcome: commenting on the data and conclusions; evaluating limitations

Summarise data patterns, trends and anomalies as revealed by your analysis of data (questionnaires, activity patterns, environmental quality). Provide a summary of the fieldwork process. What evidence in the urban area(s) is there of successful rebranding? How reliable are the findings in terms of balancing qualitative and quantitative data? (Too much of one may lead to unreliability.) What other factors could have affected results (poor weather leading to low visitor numbers, or out of season)? Environmental quality can be subjective unless pre-calibrated. How does first-hand fieldwork data compare to internet images of the urban rebrand? How could the design process be modified in terms of sample location, sample size, techniques, methods of presentation, and analytical tools and techniques?

Summary

Rebranding places

- Recognise that rebranding can incorporate both regeneration and re-imaging; marketing 'place' is very important.
- Appreciate that rebranding normally needs to tackle: social, economic, political and environmental aspects if it is to be successful.
- Know about the causes of decline in different localities: towns and cities, the countryside and coastal areas.
- Know how to design a survey to look at the profile of an area in both an urban and a rural context.
- Know how research sources can be used to explore problems of an area as well as look for evidence of managing success.
- Appreciate that the need to rebrand in rural and urban areas may be different, although there will be some overlap.
- Appreciate how creating a tourism brand may be an important part of rural and urban solutions.
- Recognise that urban and rural rebranding strategies may not always be successful or popular with different groups and stakeholders.

Questions & Answers

The specimen questions in this section are similar in style to those in the real Edexcel examination. Each question is worth **35 marks** and you will be required to answer two in the exam: one from Section A and one from Section B. You should allow about 35 minutes for each question.

Section A options are on the topics of Extreme weather or Crowded coasts. The options for Section B are Unequal spaces or Rebranding places. The structure of the questions in both sections is as follows:

- Part (a) questions involve data response, in which you are asked to use resources provided to describe or give reasons for what is shown. Up to **10 marks** are awarded.
- Part (b) questions are based around fieldwork and research activities related to topics highlighted as suitable for geographical investigation. Your response will need to show reasoning, and you should refer to your own investigations whenever possible. This part carries the higher mark allocation of **15 marks**.
- Part (c) questions will expect greater knowledge and understanding of topics, and you should use suitable case studies in your response. More detailed analysis or evaluation is required. The mark allocation here is **10 marks**.

Note that parts (b) and (c) may be reversed if this helps the structure of the question.

Sample student answers have been provided here for each question. They illustrate a mix of different responses from students. Each is accompanied by examiner's comments (indicated by the e icon), which explain where credit is due and where weaknesses or errors occur.

In the examination, answers will be marked using levels of response. The table below shows typical bands used in levels marking.

Level 1	Few points developed, showing limited explanation and often incomplete. Structure and written language weak.
Level 2	Demonstrates clear reasoning and some detail. Students could refer to actual data, places and geographical ideas. Clarity of written language.
Level 3	Shows good detail or range, often referring to concepts and case studies. Could involve some evaluation and good use of geographical terms. Well structured.

Topic 1 **Extreme weather**

Section A Question 1

Students should use the resources provided, their own ideas, and relevant fieldwork and research.

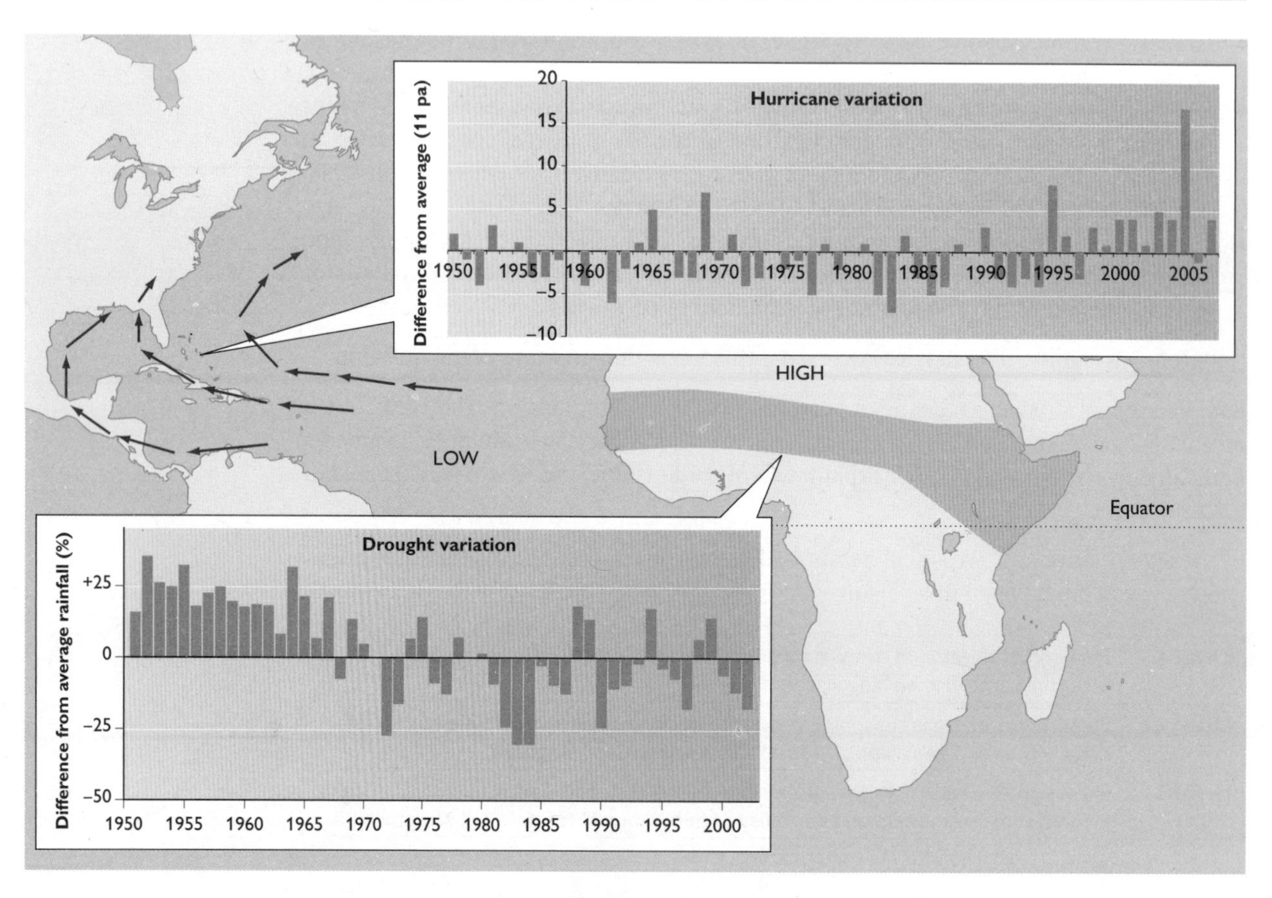

Figure 1 Two types of extreme weather

(a) Using the map and data in Figure 1 comment on the locations and patterns of extreme weather shown. (10 marks)

ⓔ It is most important to tackle both aspects of the question and to refer to the data provided.

(b) Describe and explain a programme of fieldwork and research designed to investigate the impacts of local flooding along a named stream or river. (15 marks)

e It is important to convince the examiner that research and fieldwork have been carried out by you. Detail almost always brings good marks.

(c) Evaluate the success of new technology in dealing with extreme weather events. (10 marks)

e Students whose knowledge is up to date are well rewarded, especially if they cover a range of different aspects of a topic.

Student answers

(a) Atlantic hurricanes typically travel northwest making landfall either on the US east coast or crossing into the Caribbean, before turning northeast over the US Gulf coast, like Katrina in 2005. **a** Others, like Mitch in 1998, continue even further west to strike countries like Honduras. Some never reach land. Hurricanes begin over the Atlantic in the low-pressure systems north of the equator. The warm ocean causes convection and this triggers thunderstorms. These join to make tropical storms which, in the right conditions (27°C minimum ocean temperature), grow into a hurricane. **b** The path taken by hurricanes is affected by the spin of the Earth (the Coriolis effect). The exact track each one takes can be unpredictable and can change rapidly, especially as it nears landfall.

The number of hurricanes occurring each year seems to have increased since the 1990s, changing the previous average (11) considerably. **c** The highest number was 28 in 2005. The data do not tell us how severe the hurricanes are, just how many, but 2005 did have three of the worst hurricanes on record.

The drought zone of the Sahel is in Africa, south of the Sahara. **a** Countries here suffer when the rains fail. This location, also just north of the equator but this time on land, has both a dry and a wet season. This seasonal change is caused by the ITCZ, a tropical version of the polar front, which affects UK weather. If high-pressure conditions continue, the dry season will last longer because falling air reduces rainfall, extending the drought.

The pattern of rainfall each year (in the wet season) was about 25% above average up until 1970. Since then the figures have been varied and the rain unreliable. The 1980s were especially dry (over 25% below average) and there was a famine in Ethiopia. **c**

Global warming may be one of the causes of these changes in extreme weather patterns. Some scientists think we get fewer hurricanes in El Niño periods when the circulation of the oceans oscillates (ENSO). The 2- to 3-year variations in the drought pattern may support this idea, but experts are not sure. **d**

e **10/10 marks awarded (grade A).** This is an excellent answer that deals with all aspects of the question. **a** Both locations are identified clearly. **b** These are then linked to hurricane formation and how droughts develop. **c** The patterns in the two graphs are examined and figures referred to. **d** The student has also added his/her own knowledge and understanding and has used many appropriate geographical terms. This is a well-structured account and the written language is accurate.

(b) Flood impacts can be found by fieldwork. There are sometimes markers that show previous flood levels, and the position of defence walls gives an idea of where the water came up to. One task would be to use a large map to plot the land use around the river. **a** Flood risk areas are often farmland, school playing fields and parks, rather than buildings, so that there is less damage. If you did this as a class you could put results together and cover a larger area. A transect across the floodplain would also show how flood levels and buildings are linked. Fieldwork in floods is dangerous but you can take photos from bridges and high ground to show what the flooding can do. People living near the river can be interviewed to find out how costly or damaging the floods were.

Researching floods is easier because the Environment Agency website has maps of the areas that flood and how often. **b** You can find places using postcodes. They also keep river flow records and flood heights. Local newspapers will have articles or a website about floods and how people and businesses were affected. Multimap is an online resource to use that links maps and photographs together.

We studied flooding in Keswick and plotted the land use and flood limit of the river Greta on a map. **c** We also visited and marked places where there had been damage and took our own photos to compare with those on the Westmoreland Gazette webpages, taken in the floods in 1998 and January 2005. We also talked to some local people who explained how their houses and shops were flooded

e 11/15 marks awarded (grade B/A). **a** This answer describes some fieldwork tasks, and **b** secondary sources are identified. **c** A named river is mentioned and the work carried out there is described, but not in much detail. It might have been better to discuss the tasks and the location/example together. The answer looks at fieldwork, research and then an example, but is repetitive. There is room for more geographical terminology.

(c) New technology, such as satellite monitoring, radar and computer science, has many severe weather applications. Hurricanes, tornadoes, floods and drought can all be watched and warnings given. **a**

Hurricane warning systems in the USA come from the National Weather Service and the National Hurricane Center in Miami. These rely on satellites, aircraft and buoys to track and report hurricane activity. Computer models help predict their path, wind speed and likely landfall. **b** These can be very accurate, helping to save lives. **c** However, even with this technology, there are some failures. Hurricane Mitch in 1998 changed course unexpectedly and warnings to people in Honduras and Guatemala could not save the 20,000 people who died. This also shows the importance of good telecommunications. In the USA warnings allow hundreds of thousands of people to evacuate hurricane risk areas as the storm approaches. The Met Office in the UK does a similar job with its sophisticated computer predictions and severe weather warnings.

Tornadoes are not easily predicted as they are usually small scale and they occur rapidly. But even here warning times are increasing using Doppler radar and other high-tech devices. The media storm chasers have certainly made people more aware of the risks.

Flood levels in the UK are monitored by the Environment Agency and in the USA by USGS, which both use automatic river flow gauges to collect data quickly, some transmitted by satellite. In the USA this information can be shown in real time on websites (Waterwatch) and processed by computers to predict floods and warn people automatically. There are different levels of warning up to evacuation. Modern technology is also being applied to temporary devices to keep water away from property, such as inflatable barriers and floodgates.

Drought forecasting can also be carried out using satellites remotely sensing water levels and moisture in crops and vegetation. This Drought Index is useful to farmers and water managers. Systems like this operate in the USA, Europe and Australia but the technology and data are already global. New technology does not have to be high-tech. Arguably in many poorer countries it is the introduction of new appropriate technology that helps most. **d** Organisations like Water Aid and Farm Africa are using very simple techniques, such as water harvesting and water storage, to cope with drought.

e 10/10 marks awarded (grade A). a This answer introduces a good range of severe weather types — hurricanes, floods, tornadoes and drought. **b** It explains how technology is used in forecasting and warnings and there is clear evaluation throughout the essay. **c** The point about appropriate technology is an interesting one. **d** The answer is well structured and geographical terms are used well.

Topic 2 **Crowded coasts**

Question 2

Students should use the resources provided, their own ideas, and relevant fieldwork and research.

Figure 2 Two seaside resorts compared

(a) Using the cartoon in Figure 2, comment on the changing fortunes of the two resorts shown. (10 marks)

ⓔ The key to success here is to use the information in both cartoons and link this to the idea of a resort's changing fortunes.

(b) Describe and explain a programme of fieldwork and research you would use to investigate the threat from either coastal erosion or flooding. (15 marks)

e It is important to be very direct in an investigation question focusing on fieldwork and research. Relevant and detailed evidence from real places and sources earn marks quickly. Explaining what you did is good as long as it avoids becoming a 'story'.

(c) Using named examples explain recent changes in UK coastal defence strategy. (10 marks)

e Strategy is about the approach used when protecting coastlines. It involves planning and policy rather than just a description of different types of coastal defences. Focus on ideas like SMPs, hard/soft engineering and hold the line/retreat.

Student answers

(a) Resorts go through a cycle of development. **a** They grow rapidly once large numbers of visitors can get there cheaply. For nineteenth-century British seaside resorts like the one shown, the railway was the key factor. Blackpool grew because workers from the cotton mills were able to travel there on holiday. **b** However, new destinations emerged and attracted tourists away from British resorts. This meant that many traditional resorts lost out and, despite their piers, arcades and cheap guesthouses, they began to decline in the 1960s. The result is what you see in the cartoon: run-down buildings, closed facilities and unemployment.

In the second resort the situation is different. **b** This time it is air travel that has led to the rapid growth in population, hotels and beachside facilities. This growth began in the 1960s as UK and other EU tourists opted for package holidays, with their cheap flights and family accommodation. The Blue Flag beaches, coastal scenery and sunny weather contrast with the English weather in the cartoon. This was boom time for the Mediterranean resort, bringing jobs and prosperity.

However, the cycle does not quite end there. It is possible to rebrand resorts with conference centres or theme parks and to slow the decline. Some Mediterranean resorts like Benidorm have become overcrowded, with traffic jams and high-rise hotels. Such places need to be wary of competition from new destinations, cheaper long-haul flights and new types of holidays like ecotourism. **c**

e **10/10 marks awarded (grade A).** **a** This student has used the resort cycle as a useful structure for the answer. **b** Both cartoons are identified clearly. This puts the focus on to why they are changing rather than just how they look now. **c** The student has also added his/her own knowledge and understanding and used many appropriate geographical terms.

(b) Research into the Holderness coast reveals that this is one of the fastest eroding coastlines in Europe. Old maps refer to 29 villages disappearing. The East Yorkshire Coastal Observatory website run by Hull University is a good place to start research. It uses archives of old OS maps and satellite images to plot the pattern of coastal retreat along the Holderness coast. **a** The detailed rate of erosion along the coastline (on average 2 m per year) is monitored by East Riding Council, using a series of mileposts set in the beaches. Particular storm events and the places eroding fastest are often written about in local newspapers such as the Holderness Gazette. Old photographs and OS maps are especially useful in showing where houses have been lost to the sea and just how quickly the beaches and cliffs along this coast have retreated.

Mappleton and Kilnsea are good places to carry out fieldwork. Even though defences have been built at Mappleton to protect the village and main road, evidence of recent erosion can be found. We measured and photographed the small bay that is forming beyond the rock groyne. **b** Waves have undercut the cliffs and longshore drift is scouring material from the narrow beach. Slumping has already caused the fence and part of the car park to collapse. Three km further down the coast at Cowden erosion has destroyed two farmhouses. There are very few residents in the village of Mappleton, so trying to survey local views about the erosion could be difficult — arranged interviews might be better than a questionnaire.

Near Kilnsea is the Bluebell Visitor Centre. **c** Between here and the coast are the remains of old groynes, seawalls and embankments, all unable to protect the area from erosion by winter storms. Driving onto the spit, you have to take detours to avoid places where the old road has been destroyed by the waves. **d** Further south, the line of the old railway is at an odd angle, showing how the spit is migrating as it erodes on the east and grows on the west. These features could be plotted on a copy of an old map to see the pattern and rate of erosion.

e 12/15 marks awarded (grade A). a This is generally a well-planned account, which refers to the student's own research sources and **b** fieldwork. **c** The response explains how to investigate coastal retreat/erosion at real locations. **d** However, the answer also tends to be descriptive in places – going into 'story mode' on occasion.

(c) Recent changes in coastal defence methods have seen traditional hard defences being replaced by soft and sustainable approaches. Hard defences, like seawalls and even groynes, were built at important places such as holiday resorts to protect them from erosion. These absorb the power of the waves. The problem is that if you protect one part of the coast another gets eroded or flooded. Resorts like Hastings with its pier, harbour and groynes is causing problems further east. Rock armour is now used at Fairlight and a shingle embankment and floodwall protect Pett Level from flooding. **a** These hard defences are getting old and costly to repair.

The new government policy, supported by Natural England and DEFRA, is for places where few people live or there is only farmland should be allowed to flood. This is to save money and let nature take it's course. **b** With no more maintenance and rising sea levels, places like Pett Level will soon be covered by the sea. **c** This strategy of managed retreat may not be popular with some residents but it does free money to protect more populated and important places like resorts and harbours. At Farlight new houses are still being built but further inland away from the cliffs. **d** You cannot spend more on defences than the value of what you are trying to protect.

e 8/10 marks awarded (grade B/A). a This answer is succinct and generally accurate; it shows some knowledge of a named area of coast. **b** It describes how hard defences are being replaced by a managed retreat policy and **c** begins to explain the implications of such changes. **d** Geographical terms are used but there are some written language and grammatical errors.

Topic 3 **Unequal spaces**

Question 3 Section B

Students should use the resources provided, their own ideas, and relevant fieldwork and research.

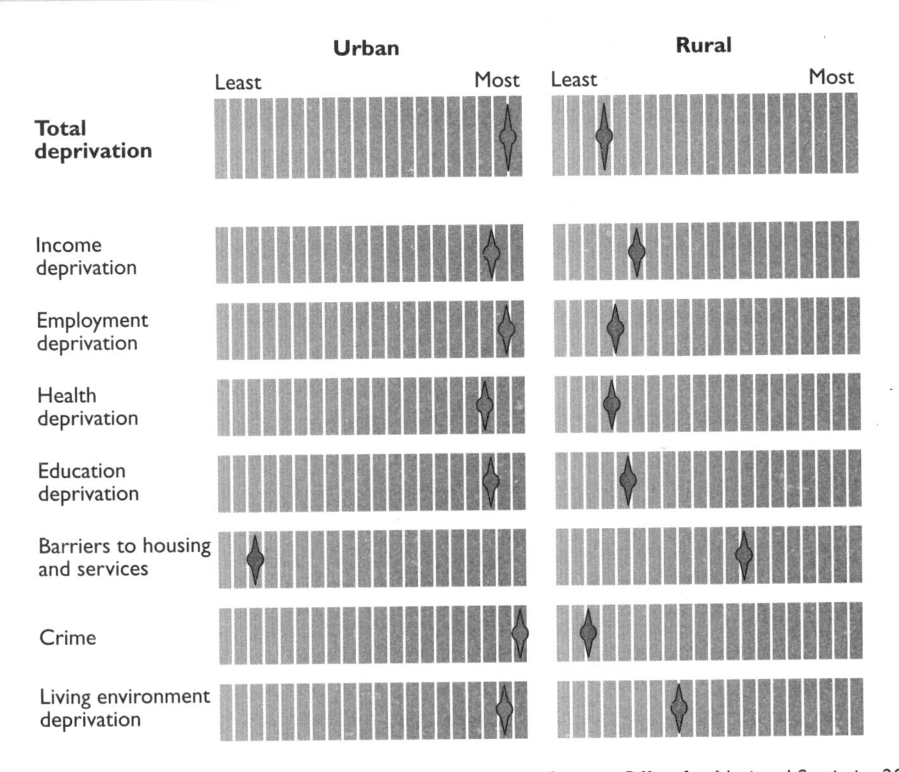

Source: Office for National Statistics 2007

Figure 3 Contrasting deprivation levels for an urban area and a rutral area

(a) Outline a range of fieldwork and research techniques that can be used to judge inequality. Comment on their relative advantages and disadvantages. (15 marks)

Be aware of the order in which the parts of the question appear – question (a) is not always data response. The big mistake here is writing about fieldwork only.

(b) Study Figure 3. Comment on the differences in the profiles. (10 marks)

It is important to make use of any data provided.

(c) With reference to *either* one named rural area *or* one named urban area, evaluate the success of different schemes to reduce inequality. (10 marks)

e Do not generalise unless you can support what you say with facts. Unfinished, short or vague answers can lose you valuable marks in the exam.

Student answers

(a) Inequality is about unevenness — the 'haves' and 'have nots'. There are several types of inequality, including spatial, economic, social, technological and environmental. **a** All of these can be investigated through primary fieldwork using a range of strategies. Fieldwork relating to inequality usually involves a comparison of two or more areas.

To look at the general inequality of an area, an important tool is the quality survey. Often these are bi-polar scores. This approach can be customised to fit a range of locations and objectives. It can also be supported with digital photographs. It has the advantage of being relatively quick and easy to set up, and is also straightforward to carry out in the field (no specialist equipment or training is required). One of the big problems with the approach, however, is that the results can often be completed in too hurried a manner without proper thought and attention to detail. This can be overcome to some extent by 'pre-calibration' of sheets as a group exercise. **b**

Environmental inequality can also be surveyed using dedicated bi-polar sheets, but in addition to this, litter and pollution (e.g. air quality) surveys may also be applicable. Litter surveys work well if you use a systematic approach and note down the amount and type in dedicated areas. Air quality can be monitored (e.g. lichens), but quantitative approaches are difficult and expensive. **c**

Questionnaires and interviews can also be used to judge inequality. These have the advantage that they can give the views of respondents, while being easy to deliver in the street. Their disadvantages are that not everyone wants to answer (issues with fairness and sampling) and that questionnaires are difficult to design.

More innovative approaches to social inequality may be to use graffiti assessment indexes, or alternatively to study accessibility in the built environment, e.g. measuring the gradient of pavements and the width of footpaths.

e **9/15 marks awarded (grade C/B). a** There is a good and clear introduction here, showing thorough understanding of the topic. **b** This answer covers a good range of fieldwork ideas, all of which are relevant to inequality. **c** They are well supported by advantages and disadvantages. However, there is no mention of any research so this answer is likely to be limited to a maximum of 10 marks. It could also have been improved with reference to a real place and the student's own fieldwork.

(b) According to the diagram, deprivation is composed of a number of elements: income, employment, health, education, housing/services, crime and living environment. The scale shows a composite of all of these factors, which we have to assume are not weighted in any way.

Both profiles are very different in terms of their overall deprivation: the rural area shows much less deprivation that the urban area. In general, the rural area has lower levels of deprivation for many of the factors, including income, employment, health, education, crime and living environment. Barriers to housing/services in the rural area, however, show significant deprivation. **a** This could be due to low affordability potential (i.e. high house prices — the 'rural idyll' — combined with lower average salaries/seasonal work). There is a lower density of services in the rural area and they take longer to reach.

Crime shows significant deprivation in the urban area (it has the 'worst' score). This compares with little deprivation in the rural area. Differences in crime can be linked to population density and perhaps higher rates of reporting in towns than in rural areas. **b**

Employment deprivation also shows high contrasts. Again, cities have a large range of jobs types; these data indicate that there are poorly paid jobs in this particular urban area (this may not be the same for all urban places). The rural area has lower levels of deprivation in terms of both income and jobs.

I suspect the data are based on small geographical areas. They may have been selected to show maximum contrasts and are not necessarily reflective of all urban or rural areas. **c**

e 10/10 marks awarded (grade A). a A very clear, logical and well-structured response. It shows both range and depth of detail in terms of the coverage of the factors, with some sensible reasons identified. **b** The command words 'comment on' require elements of both description and explanation, which this response has achieved. **c** The last statement is particularly insightful — recognising that the data have probably been chosen to show maximum differences. Good spot!

(c) I am going to be talking about schemes to tackle rural inequality in Cornwall. West Cornwall gets Objective 1 funding from the EU as it is highly deprived. **a** This means it has high levels of unemployment, low standards of education, poor services etc. The EU funding is designed to help develop infrastructure, education and training. **b**

West Cornwall Together is a partnership involving lots of individuals and groups of people. People come together to try and improve quality of life and standards of living. There are other aims such as environmental protection, healthy life for all residents and jobs for people. **c**

Another group of people trying to reduce inequality are the community regeneration team, who are part of the local council. They support projects such as play schemes and give away grants, including the provision of ICT in community halls and farmers' markets.

e 6/10 marks awarded (grade C). a A reasonable start, but there are some sweeping generalisations in the first paragraph — not all parts of west Cornwall are deprived. **b** It was a good idea to talk about what the funding is designed to do. **c** The section on West Cornwall Together needs more depth and detail. Overall the answer is acceptable and shows reasonable structure. It would have been improved by a bit more depth and a few more facts and figures. Success was not discussed which would probably limit the marks to a maximum 6 out of 10 as the question asks for evaluation.

Topic 4 **Rebranding places**

Question 4

Students should use the resources provided, their own ideas, and relevant fieldwork and research.

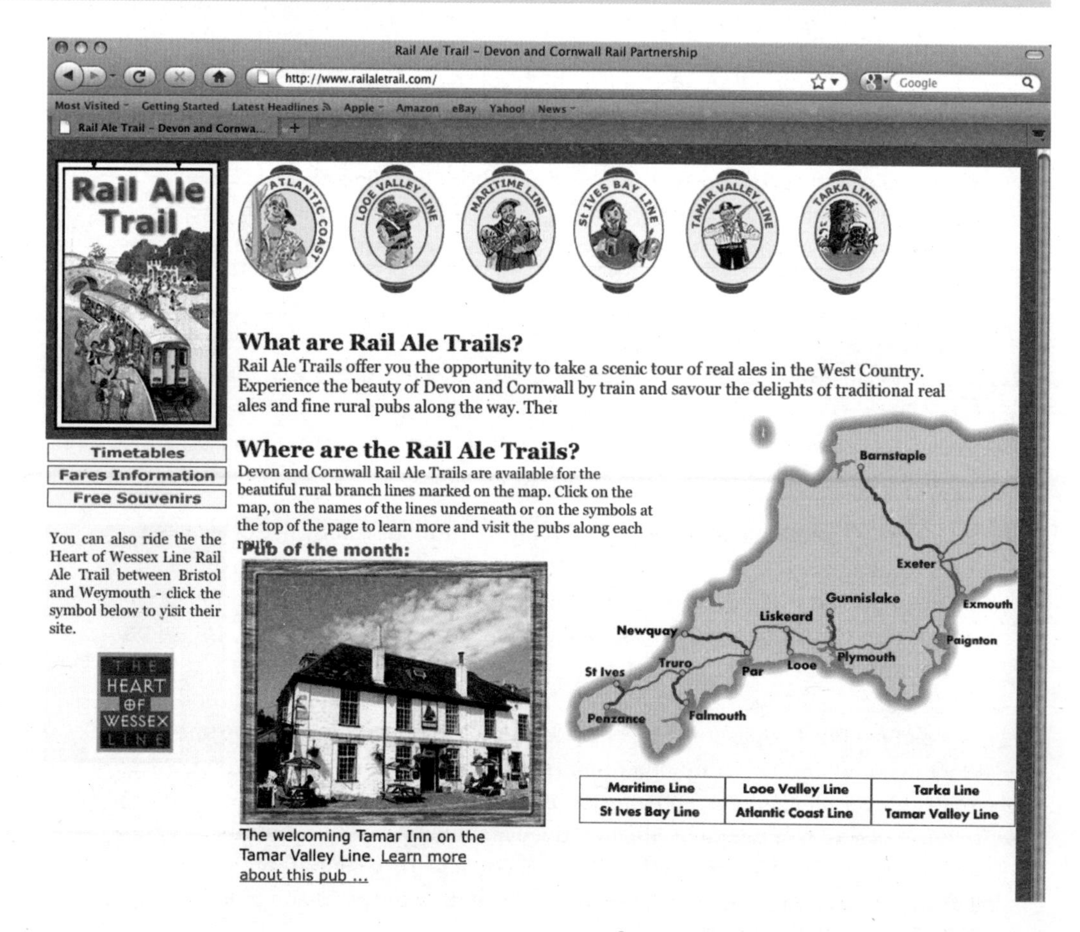

Source: www.railaletrail.com/wessex/index.htm

Figure 4 The Rail Ale Trail website

(a) Using Figure 4 suggest how 'image' can be used to promote rural tourism. You should illustrate your answer with reference to named examples. (10 marks)

Try to plan your answer if you can – that way you will not lose focus. A well-argued answer does not have to be filled with 'loads of facts', just those which matter.

(b) Explain how you would plan a fieldwork and research exercise to determine the profile of an area in need of rebranding. (15 marks)

e Fieldwork and research activity need to be convincing. Knowing why tasks are done and which research is useful is really important. 'Taking photos' and 'visiting websites' are not enough. Get involved!

(c) With reference to named examples, outline the reasons why rebranding is needed in either urban or rural places. (10 marks)

e Using precise and accurate terminology delivers marks quickly. Terms like 'deindustrialisation', 're-imaging', 'dependence', 'challenging circumstances' and 'catalyst' show a high level of understanding.

Student answers

(a) Images selectively portray people and places. In that sense they can often be misleading or used to market a particular product or service. Since images are selective, what is most often interesting is the negative information which might be missed out. **a**

'Images' can take a number of forms including websites, poems, photographs, videos/DVDs or paintings. They can be in the visual, written or spoken form. In Figure 4, the image is a website which has an important embedded cartoon which is used to promote rural tourism. The use of local beer is a clever marketing tool, since it sells ideas of 'wholesomeness', a 'fun day out' and also local distinctiveness. Another advantage of the rail ale scheme is that it encourages visitors to use the train. This is a mode of transport that they may not have thought about using (they would have probably used their own cars) so it may also encourage the use of rural branch lines at off-peak periods, which is good for the train company.

I carried out a fieldwork and research exercise in Surrey and saw similar examples of how local distinctiveness was marketed through images (leaflets, posters and websites), thereby attracting visitors. One notable example was the Denbie Vineyard near Dorking offering an 'indoor and outdoor' wine experience. **b** The image of idyllic countryside was being promoted through the website and fliers. **c** They had also added value to the facility by making available accommodation and facilities for conferences.

e **10/10 marks awarded (grade A).** **a** A clear and well thought out response that is focused on the question and gets to grips with how 'image' can be used. **b** It demonstrates clear understanding and uses a good example. **c** The student has also added his/her own knowledge and understanding and used many appropriate geographical terms. This is a well-planned account and the written language is accurate.

(b) For our fieldwork we went to Shrewsbury to see how it could be rebranded. **a** Part of this included looking at what Shrewsbury is like at the moment. We were divided into six groups and had a part of the town to look at and investigate. **b** We were given different recording sheets, including environmental quality and had to go around at look at our section of the town. I also took photographs, even though it was raining.

We planned the exercise so that the town was divided into six roughly equal units. **c** We also designed recording sheets to investigate the profile of the town.

These included spaces to note down gut feelings and other general observations. We also used the internet to find out what people thought of the town and its current image. This meant I visited various websites and the local internet blog. **d** I found that that people generally liked the town, although they did tell us about parts of the town where they thought it could do with improving.

6/15 marks awarded (grade D/E). **a** This is generally a narrative of the fieldwork day rather than a focus on the question. **b** There is no real feeling that the student has had much personal involvement in the design process, **c** although there is evidence that he/she helped during the planning stages. More emphasis is required on a discussion of the idea of a 'profile'. **d** Research is thin and weakly described. Reference to specific websites and/or reports would have been a big improvement. It is a shame there wasn't more focus on the idea of 'profile', which was central to this question.

(c) Rebranding places has become increasingly important, particularly in areas which have suffered economic and social decline as the result of deindustrialisation. Rebranding can also include ideas about re-making, re-imaging and regenerating. In essence it has an environmental, social, political and economic focus of 'improvement'. **a**

Coastal areas are a good example of a particular location that has needed to rebrand. Towns and resorts at the coast face particular problems, not least because of their remote geographical access. They may also be plagued by their dependence on a shrinking and seasonal resort economy. **b** On a recent field visit to Great Yarmouth in Norfolk, we found evidence (via interviews) of the problems of emigration and immigration, especially conflicts between some residents and eastern European economic migrants. The town also exhibited run-down parts and physical decay and was in need of a rebrand. **c**

Blackpool is another example of a coastal resort that has faced challenging circumstances. In the 1980s there were 17 million visitors annually, but now there are about 10 million. Fewer tourists have resulted in fewer jobs and the number of businesses fell by 6% in the period 1994–2005. A significant number of districts in the town are in the top 10% for deprivation nationally. Blackpool wanted to use a super-casino as a catalyst for rebranding, but the bid was rejected by the government. It will now have to come up with a new scheme to reinvent itself. **d**

10/10 marks awarded (grade A). **a** A very good start with a clear and well-presented interpretation of the process of rebranding. **b** Coastal areas are a good choice since they have a number of unique problems that can be discussed easily. **c** Good use is made of the student's own fieldwork and research in this response. **d** The answer also has well developed exemplary material based on Blackpool. Many geographical terms are well used.

Knowledge check answers

Section A

1 Weather that is either severe or unexpected.

2 Strong winds, torrential rain, massive waves, floods.

3 They begin along the ITCZ north of the equator, sea temperatures of over 26 °C and air humidity over 75% trigger tropical storms. They form a spiral of increasing wind speeds that bring torrential rain. The Earth's rotation (Coriolis effect) means hurricanes migrate northwestwards towards the Caribbean and Atlantic coast of the USA.

4 Hurricanes occur from July to October due to high sea temperatures; droughts in summer/dry season as a result of anticyclones (blocking highs). Depressions happen mainly in autumn and winter over the UK because of Atlantic airmasses and low pressure.

5 The way weather features change over time, the pattern of weather systems and features (cloud, fronts, etc.) and the role of pressure systems and air masses.

6 It arranges hurricanes into five categories based on wind speeds over 119 km per hour.

7 You could plot land use, show extent of flooding and use them to carry out transects.

8 The likelihood of a flood of a certain size/level returning in a certain time.

9 Hydrographs, flood recurrence interval graph, photos before and after, GIS mapping, etc.

10 Allow unused areas to flood to save others and reorganise land use to reduce damage.

11 No correct answer but the most sustainable responses are – set aside, afforestation, restoration, washlands.

12 See and track hurricanes; improve storm and flood prediction; give warnings and organise evacuation.

13 Using simple solutions like bunds, water carts or wells; developing drought-resistant crops; collecting rainwater.

14 Sampling issues, bias and out-of-date information.

15 Relates to (a) employment, natural resources, accessibility and history and (b) development of hotels, facilities, and increasing population.

16 Better weather, cleaner beaches, package holidays and cheap air travel.

17 The effects of pollution; damage to ecosystems by trampling, etc.; the views of developers, residents, visitors and interested parties.

18 The effects that development has on ecology, people and their lives; a financial calculation of gains divided by costs (at least +1 or better); the likelihood of negative effects.

19 (a) Fieldwork = old buildings and roads; damaged coastal defences; cliff slumping and rock falls; longshore drift. (b) Sources = Environment Agency; university research; a local newspaper; historical records/photos; GIS mapping (Google Earth).

20 Hard engineering uses artificial structures to prevent erosion/flooding and soft engineering uses natural systems (ecosystems) to defend coasts.

21 Groynes are low cost, look OK but can cause erosion downdrift. Sea walls are costly to build and maintain but useful in high-value locations.

22 Nourishment = sand is pumped/brought in to help resist beach erosion. Retreat = prevents new building but compensates residents at risk.

23 The more you spend the longer they last.

24 Realignment allows unused areas to flood or erode to save others, using e.g. salt marshes. SMPs integrate various techniques to tackle a range of erosion/flooding problems along a coast.

Section B

25 Spatial inequality is the unequal amounts or qualities of resources and services depending on the area or location, such as medical care or access to employment.

26 Focus groups are a type of qualitative research in which a group of people (perhaps 5–10) are asked about their perceptions, opinions, beliefs and attitudes towards a product, service, concept, development, etc. Interviews tend to be formalised and semi-structured; normally they are on a one-to-one basis.

27 From the Office for National Statistics (ONS): 'Geographical areas within settlements of over 10,000 people are defined as urban, whilst those in settlements of under 10,000 people, are rural settlements, including: Town and fringe, Village, or Hamlet and isolated dwellings.'

28 Households where income is less than 60% of median household disposable income (ONS).

29 Survey the main streets in an area with different categories, e.g. percentage vacant, high order shops, etc. This could allow the build-up of a specialised land-use map. Also, 'clone town surveys' might be a useful indicator of quality.

30 Transforming a farm into another enterprise to make money, e.g. holiday cottages, campsites etc., rather than producing crops in the traditional way. It is often associated with farm diversification.

31 IMDs are a common way of aggregating several indices of deprivation (education, incomes, etc.) into one single measure.

32 Different groups of people (based on ethnicity) living in different areas, often physically separated and not choosing/being able to mix.

33 Cultural rebranding is often based around the promotion of culture and the arts, e.g. cinemas, galleries, museums etc., whereas heritage is more about the historical significance of a location.

34 This refers to government action taken to stimulate the economy of an area through the introduction of public sector work and jobs.

35 An economy based on knowledge, ideas and services, often called the 'new economy'. Relies heavily on interconnections through the internet.

36 A high profile and generally impressive (sometimes controversial) new architectural build that may form the flagship to a new rebranding scheme. Typically bold design with modern materials.

37 Usually schemes that kick start rebranding in other areas. May be government funded. They make the area attractive for others outside the area and for new investment.

38 Oral history = someone's narrative about a place, usually its history. Retail diversity = a measure of how many different types of shops there are. Normally high diversities are associated with a successful shopping environment.

A

B

C

D

E

F

G

H

I

L

M

N

O

P

R

S

T

U

V

W

Y

ISBN 9781444147643